NEW ZEALAND WETLAND BIRDS
and their World

NEW ZEALAND WETLAND BIRDS
and their World

GEOFF MOON

NH NEW HOLLAND

First published in 2009 by New Holland Publishers (NZ) Ltd
Auckland • Sydney • London • Cape Town

www.newhollandpublishers.co.nz

218 Lake Road, Northcote, Auckland 0627, New Zealand
Unit 1, 66 Gibbes Street, Chatswood, NSW 2067, Australia
86–88 Edgware Road, London W2 2EA, United Kingdom
80 McKenzie Street, Cape Town 8001, South Africa

Copyright © 2009 in text: Geoff Moon
Copyright © 2009 in photography: Geoff Moon except where otherwise credited
Copyright © 2009 New Holland Publishers (NZ) Ltd

Publishing manager: Matt Turner
Editor: Brian O'Flaherty
Design: Dexter Fry

National Library of New Zealand Cataloguing-in-Publication Data

Moon, Geoff.
New Zealand wetland birds and their world / text and
photography by Geoff Moon.
Includes bibliographical references and index.
ISBN 978-1-86966-197-7
1. Wetland birds—New Zealand. I. Title.
598.17680993—dc 22

10 9 8 7 6 5 4 3 2 1

Colour reproduction by Pica Digital Pte Ltd., Singapore
Printed in China by SNP Leefung on paper that has been sourced from sustainable forests.

All rights reserved. No part of this publication may be reproduced, stored in a retrieval system, or transmitted in any form or by any means, electronic, mechanical, photocopying, recording or otherwise, without the prior permission of the publishers and copyright holders.

While every care has been taken to ensure the information contained in this book is as accurate as possible, the authors and publishers can accept no responsibility for any loss, injury or inconvenience sustained by any person using the advice contained herein.

Front cover: Australasian pied stilt.
Title pages: Matata Lagoon, Bay of Plenty.
Facing page: black-fronted dotterel.
Contents page: black swan.
Back cover (l–r): white heron; Te Henga wetlands in the Waitakere Ranges; New Zealand kingfisher.

ACKNOWLEDGEMENTS

In recent years there has been an increasing awareness of the importance of our diminishing wetlands, and of the valuable fauna and flora that inhabit them. Accordingly, Belinda Cooke and Matt Turner of New Holland Publishers have encouraged me to write this book, focusing on the numerous species that rely on New Zealand freshwater wetlands for food, shelter and nesting.

In addition to the staff at New Holland, and editor Brian O'Flaherty, I want to thank the numerous friends who have helped me over the years with information and opportunities for photography.

I would particularly like to thank my wife, Lynnette, for her interest and considerable involvement in the completion of this book.

I have attempted to describe some of my own observations of the birds, especially the most secretive species, such as the bittern, about which there appears to be limited in-depth information.

Geoff Moon

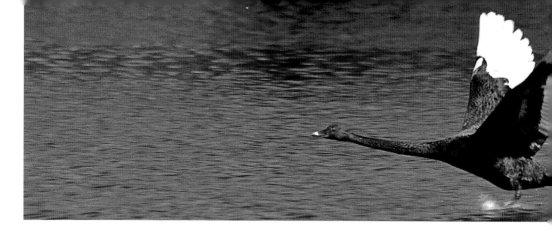

CONTENTS

Acknowledgements..5

Introduction..8

Order PODICIPEDIFORMES: **grebes**
    Australasian crested grebe....................................24
    New Zealand dabchick..27
    Australian little grebe..34

Order PELECANIFORMES: **shags**
    Black shag..36
    Pied shag...38
    Little black shag...42
    Little shag...44

Order CICONIIFORMES: **herons**
    White-faced heron...48
    White heron..55
    Australasian bittern...58
    Royal spoonbill..64

Order ANSERIFORMES: **swans, ducks, geese**
    Mute swan..68
    Black swan...70
    Canada goose...72
    Paradise shelduck..74
    Blue duck..76
    Mallard..80
    Grey duck...82
    Grey teal...83
    Brown teal..84
    New Zealand shoveler..86
    New Zealand scaup...86

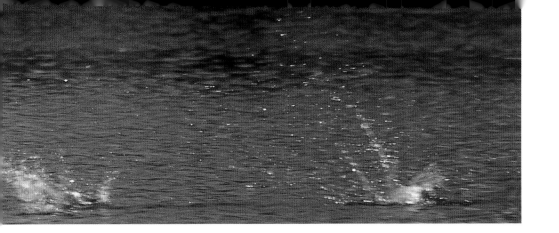

Order FALCONIFORMES: **diurnal birds of prey**
    Australasian harrier..88

Order GRUIFORMES: **rails, swamphens and coots**
    Banded rail..92
    Spotless crake...95
    Marsh crake..98
    Pukeko..100
    Australian coot...101

Order CHARADRIIFORMES: **waders, gulls and terns**
    South Island pied oystercatcher...104
    Australasian pied stilt..106
    Black stilt...110
    Banded dotterel...112
    Black-fronted dotterel..115
    Wrybill...116
    Spur-winged plover...118
    Black-billed gull...121
    Black-fronted tern..123

Order CORACIIFORMES: **kingfishers**
    New Zealand kingfisher...126

Order PASSERIFORMES: **passerine birds**
    Welcome swallow...134
    Fernbird..136

    Glossary..140

    Bibliography...141

    Index...142

INTRODUCTION

'O let them be left, wildness and wet;
Long live the weeds and wilderness yet.'

– GERARD MANLEY HOPKINS

For many years naturalists have appreciated the importance of wetlands as complete ecological systems, voicing their concerns when these locations were drained and converted into agricultural land. In recent years, however, there has been a growing awareness of their importance as essential habitats, not only for particular bird populations, but also for diverse species of fish, insects and other invertebrates and the unique plants found only in wetlands.

Focus on freshwater bird species

This book focuses on the bird species that essentially favour freshwater wetlands. Of particular importance are the birds that depend totally on the survival of inland freshwater habitats. Some wetland birds included in the book, such as kingfishers, readily fly to varying water locations, including the coast. But others, notably the fernbird, rely on specific freshwater domains. These birds depend not only on the food source found in these particular wetlands but also the vegetation they support.

Our wetlands provide permanent residence for many secretive bird species, while others visit them to feed or nest. In fact, well over 50 of our bird species use wetlands at some time.

Dune lakes and shallow lagoons provide food and shelter for wetland birds. The North Island dune lakes support good populations of New Zealand dabchicks and Australian little grebes, as well as the secretive bittern and many waterfowl.

In 1975 scientists from several countries who were alarmed at the worldwide destruction of wetlands met in Ramsar, Iran to establish a convention on wetlands. The aim of the convention agreement was to list and register significant wetlands for permanent protection and enhancement. New Zealand is now a party to the convention, and several of our country's wetlands have been listed as Ramsar sites.

Today, New Zealand has only about 10 per cent of its original wetlands, and although many

The Vernon Lagoons in Marlborough. Like the much larger Lake Ellesmere in Canterbury, these two shallow lakes are isolated from the sea by boulder and shingle banks. They provide prime habitats for very large numbers of wetland birds.

of these areas are now protected, some, especially a few of our larger lakes, have their water nutrient enriched by run-off from adjacent farmland. This causes excessive growth of algae and an overabundance of waterweed. In an attempt to counter this, efforts are being made to encourage farmers to fence off strips of land bordering streams that feed into lakes and to control animal effluent outflow.

Willows introduced in the 1880s are well established along the banks of numerous lakes and rivers, replacing native vegetation and also causing problems by altering the flow of many streams. Some birds have adapted to this environment, but others have retreated to swamplands of raupo and rushes.

Wetland types

For nesting, food or shelter, the wetlands most favoured by a large variety of birds are **swamps** that provide extensive growths of raupo and other vegetation, interspersed with expanses of open water. Shallow **coastal lagoons**, **dune lakes** and **braided rivers** are other ideal habitats. Many **bogs**, with their sponge-like character, are home to fish and insects but are not usually favoured by birds. Farm **drainage ditches** and

The Te Henga wetland in the northern Waitakere Ranges is typical of bittern, crake and fernbird habitats, with dense stands of raupo and stretches of open water.

small ponds are also useful to a few species of birds. In particular, little shags are often seen feeding in drainage ditches.

There are many **large lakes** in New Zealand. The largest is Lake Taupo, which is of volcanic origin, like many of the lakes in the central North Island. Earthquakes that have uplifted land to block off valleys have formed some lakes. Lake Waikaremoana is an example of this.

Some South Island lakes were carved out by glacial action during the last ice age. As a result of this, many of them are very deep and clear, features that particularly suit our only true diving duck, the New Zealand scaup. A few of these lakes are turquoise in colour, while others have a milky appearance due to the presence of rock flour, suspended particles ground from alpine rocks by glacial action.

A number of the larger lakes, including some hydro lakes, are subject to wave action, which prevents the establishment of thick vegetation on their borders. These are not favourable habitats for many species as they lack secure roosting and nesting sites. However, black swans, Canada geese and a few species of duck often inhabit some of them. These birds find the lakes useful as feeding sites but readily fly off in flocks to other lakes for roosting.

Much of Lake Taupo is affected by wave action, but where borders of the lake are

Many high country lakes are home to crested grebes, and New Zealand scaup (above) frequent subalpine lakes such as this one (left) near Lake Ohau in Canterbury.

sheltered from the prevailing south-west winds, large stands of raupo flourish and these sustain thriving populations of fernbirds, crakes and bittern. New Zealand dabchicks also favour these more sheltered enclosures along with shags, black swans and ducks.

Many of the **man-made lakes** are long and narrow, having been formed by the damming of rivers to provide water storage for generating electricity. Some of the older hydro lakes such as Lake Whakamaru on the Waikato River have built up valuable stands of raupo and other vegetation along their borders, and these habitats support populations of bittern, shags, crakes and fernbirds. The smaller hydro lakes are subject to frequent fluctuating water levels, and this prevents adequate growth of vegetative cover on their banks. However, shags, being strong fliers, frequently visit these lakes to feed.

By preventing drainage, wind-blown sand has formed many **dune lakes**, particularly in Northland, mostly along the west coast. These lakes receive water from rain, and in summer months some of the smaller lakes dry up. But the larger dune lakes are deeper and bordered with dense growths of raupo and reeds. As a result

Previous pages: Lake Whakamaru, one of the older hydro lakes on the Waikato River, is bordered by thick vegetation, making it a suitable habitat for wetland birds.

Above: Lake McGregor near Tekapo supports nesting Australasian crested grebes and New Zealand scaup.

they sustain a very large number of wetland bird species, and in particular New Zealand dabchicks. Inhabitants of some of these lakes also include good populations of the recently self-introduced Australian little grebe.

The **subalpine lakes** of the South Island are the remaining strongholds of the Australasian crested grebe, which no longer inhabits the North Island. Many of these lakes freeze over in winter and the birds move to Lake Forsyth, near Banks Peninsula, and other lowland lakes.

The large, shallow Lake Ellesmere in Canterbury and the brackish Vernon Lagoons in Marlborough have been isolated from the sea by a boulder bank in the latter, and by a large shingle spit in the former. Both locations support very high numbers of wetland birds, particularly black swans and other waterfowl. In 1978 royal spoonbills nested in a loose colony on one of the low-lying islands in the Vernon Lagoons, and in recent years they have extended their nesting sites to other coastal locations. They had previously nested only in small numbers in trees adjacent to the white heron nesting colony near Okarito in Westland.

New Zealand's wetland habitats include numerous **rivers**. Many of those in the North Island radiate from the central high country. In the South Island some of the rivers that are the most important for birdlife are the shingle

braided rivers flowing east from the Southern Alps. There are also several smaller shingle rivers in the southern North Island that are valuable bird habitats.

The vast South Island braided rivers are of major ecological importance to a number of wetland bird species. New Zealand is unique in the world for its braided river expanses, which support many species that make the river their main habitat. Typical of the unique birds feeding and nesting on braided rivers is the wrybill, named the 'crook-billed plover' by Walter Buller. With the end of its bill twisted to the right, the bird is able to forage under rounded river stones for insects and grubs. The wrybill's cryptic colour exactly matches the grey river stones in its habitat and, unless the bird is moving, it is hard to spot. Chicks and eggs are well camouflaged, hiding them from patrolling black-backed gulls or harriers.

The Vernon Lagoons in Marlborough: stands of raupo and other marginal vegetation provide a nesting habitat for many species of wetland birds.

Left: The Whirinaki River in the Ureweras is a blue duck habitat.

Above: A blue duck rests on a rock in calm water in Mt Cook National Park.

Other endemic species that nest on braided and shingle rivers include the black-billed gull, the black-fronted tern and the banded dotterel. South Island pied oystercatchers, pied stilts and recently the self-introduced Australian black-fronted dotterel also occasionally nest on these riverbeds. The black-fronted dotterel was first discovered nesting on shingle rivers in Hawke's Bay, but this bird has since expanded its range.

Most braided rivers rise in alpine altitudes, carrying rocky debris from eroding mountains and depositing it to create wide gravel waterways. The rivers are characterised by multiple interweaving channels that begin with water pouring in torrents from high rapids and flowing on to form wide open stretches of water, deep pools, streams, shallows and islands. These differing water level habitats suit various species of birds.

Periodic natural floods in braided rivers scour the gravels free of invasive weeds. Any irrigation schemes and damming of the rivers affect water flow and the regular flush of weed growth by flooding. This results instead in shallow riverbeds and a consequent spread of weed vegetation, including lupins, broom and willow.

The presence of this vegetation creates a major problem for nesting birds as, apart from restricting and diverting the natural flow of their river habitat, the growth also provides cover for stalking predators. It has recently been found that feral cats are preying on nesting black-billed gulls and black-fronted terns at night. With any water drainage, and consequent shallow river water, predators find easy access to islands where birds also nest.

Many rivers have been seriously modified by the harnessing of their waters for hydroelectric power schemes. The establishment of the Lake Benmore power scheme, the largest in the country, has most drastically affected the rivers of the Waitaki catchment and consequently the habitat of the highly endangered black stilt.

The world's rarest wader, the endemic black stilt nests and feeds only in the wetlands of the Waitaki catchment, so the drainage of the birds' habitat has denied them their usual safe nesting

Right: The Rakaia River, Canterbury. Braided rivers are important feeding and nesting environments for numbers of bird species.

Below: The wrybill is the only bird in the world which has the end of its bill twisted to the right.

A number of braided rivers have been invaded by exotic willows and weeds, making them unsuitable for some species of nesting birds.

sites and ready availability of food. They are also affected by the increase in predators that originally preyed on a multitude of rabbits in the region. Since rabbit numbers have been controlled, predators have instead found birds to be easy prey, especially when they are nesting.

Although wetland habitats will probably continue to be drained for agricultural use, there is now a much greater awareness of their value and a keen interest in the wildlife they support. Enlightened farmers are forming larger farm dams that accommodate the creation of small islands and many are planting wetland vegetation to establish secure environments. These developments have quickly attracted varieties of birdlife.

Species not covered

Not discussed in this book are the numerous northern hemisphere migrant waders and coastal marine waders, notably godwits, that visit some inland wetlands, particularly Lake Ellesmere in Canterbury. These birds primarily favour sheltered harbours and coastal estuaries.

A few other familiar species are also excluded, such as black-backed gulls, red-billed gulls and Caspian terns. Black-backed gulls, although usually inhabiting marine coastal habitats, may also be found far inland, often cruising along braided rivers to prey on eggs or chicks of river-nesting birds. (I was once photographing from a hide on the shores of a coastal lagoon when I saw a black-backed gull land on the water carrying an egg. With binoculars I identified it as a dotterel egg. The bird twice dropped the egg from a height until it broke and was then consumed.) Similarly, the ubiquitous red-billed gull is seen on inland wetlands, particularly lakesides. The large cosmopolitan Caspian tern sometimes flies far inland, and I have seen birds fishing on small lakes near Lake Taupo.

Order Podicipediformes

Grebes

Three resident breeding grebes inhabit New Zealand. These are the Australasian crested grebe, which is found only in the South Island, the endemic New Zealand dabchick, found only in the North Island, and the recently self-introduced Australian little grebe which is becoming increasingly common in the North Island with a few in the east of the South Island.

Although there were sightings of the hoary-headed grebe in the 1970s, this species failed to become established.

A particular charactcristic of grebes is that their legs are placed well back on the body, with strong muscles attached to the tibia (shin bone) of each leg. In addition, their claws are flattened and the toes are joined by webs (lobe-webbed), unlike the completely webbed feet of many other water birds, such as ducks.

Grebes are seldom seen flying during daylight hours, but are obviously strong fliers, as they move to other wetland habitats, presumably at night.

AUSTRALASIAN CRESTED GREBE
Podiceps cristatus australis

Threatened • Protected • Length: 50 cm • Maori name: Puteketeke • Status: Native

Above: Australasian crested grebes inhabit South Island subalpine lakes.

Right: A crested grebe incubating eggs.

This subspecies of the cosmopolitan crested grebe is confined to the South Island where it inhabits many of the clear, shallow subalpine lakes east of the Southern Alps. Small numbers are also found in some lowland lakes in Westland, and a few in Fiordland, although a small population that used to occupy the lakes in the upper Eglinton Valley no longer exists.

In severe winters when the subalpine lakes freeze, most birds migrate to lowland lakes such as the coastal Lake Forsyth near Banks Peninsula.

Crested grebes, which are entirely aquatic, live on a diet of small fish, crustaceans and other small invertebrates. The birds catch their food by diving neatly from the water surface and usually remaining submerged for several seconds. The longest dive I have recorded was 53 seconds.

In the early spring months crested grebes pair and spend considerable time in courtship displays, with the male bird first flourishing his upper-neck ruff and head crest to the female. Then both birds rise in the water and, facing each other, shake their erected head crests while uttering low, moaning cries.

Nesting usually begins in early October, with the paired birds building a bulky nest composed of sticks, rushes and waterweeds. The structure is anchored to trailing willow branches or raupo stems on the edge of, or growing within, a lake. A clutch of up to six white eggs is laid in November, but in cases where eggs have been robbed or destroyed in storms, a smaller second clutch will often be laid as late as January.

Top left: A crested grebe uncovering eggs before settling to incubate.

Bottom left: The sitting bird erected its head ruffs as a threat to a black swan which came too close to the nest.

Above: The crested grebe courtship display.

Incubation is shared by the sexes and eggs soon become stained with waterweed, as, like other grebes, the birds cover the eggs to camouflage them when leaving the nest. As incubation usually starts after the laying of the first or second egg, chicks tend to hatch at intervals of several days. Normally, hatchlings remain in the nest, but on one occasion I observed that when the birds had a large clutch, two of the older chicks left the nest, riding on the back of one parent while the other continued to incubate.

When all the chicks have hatched, two of them may ride on one parent's back while the other adult dives for food to feed the chicks.

As crested grebes feed their chicks on fish, the parents will often pluck feathers from their own breasts and feed these wetted feathers to the chicks. This is thought to be a method of protecting the chick's gut from trauma caused by sharp fish bones.

Chicks soon start to make shallow dives to feed themselves, although their parents continue to feed them, even until they are several weeks old. I have sometimes seen older chicks make futile attempts to ride on a parent's back.

NEW ZEALAND DABCHICK
Poliocephalus rufopectus

Threatened • Protected • Length: 28 cm • Maori name: Weweia • Status: Endemic

New Zealand dabchicks are now found only in the North Island, and are located on various lagoons, large clear-water lakes and many large farm dams. Dune lakes on the west coast of the North Island are a favourite habitat. They occasionally inhabit oxidation ponds, and even some urban recreational lakes.

As with other grebes they are seldom seen flying during daylight hours, except sometimes during courtship when skimming across the surface in a chase. However, they are obviously strong fliers, migrating to other bodies of water at night. A few years ago, dabchicks migrated to the Western Springs lakes in urban Auckland, having flown at night across vast built-up areas.

Although the New Zealand dabchick is not now recorded as inhabiting the South Island, I was with companions when we saw one on the small Lake Wombat south of Franz Josef Glacier

A New Zealand dabchick with chicks.

Left: New Zealand dabchicks now only inhabit North Island wetlands.

Above: New Zealand dabchicks' courtship display.

Right: Dabchicks copulating.

Bottom Right: A dabchick carrying nesting material.

in 1952. It approached us within 5 metres when I splashed the water with a stick. It is possible that the bird had a mate on a nest nearby. I was unable to return to check the sighting, but I reported this record.

New Zealand dabchicks are entirely aquatic in habit, with legs set well back on their bodies. Their lobe-webbed feet with flattened toes and claws allow the birds to be extremely efficient swimmers and divers.

When diving to feed, they submerge with scarcely a splash to catch aquatic invertebrates, small fish and crustaceans. The birds remain submerged for a maximum of 25 seconds, often surfacing several metres away from the point of their dive.

Right: New Zealand dabchicks' nest in a tyre, hanging from a wharf.

Clockwise from opposite top left: A dabchick approaches the nest to take its turn at incubating.

The dabchick impatiently jumping on its mate's back, to dislodge the sitting bird.

The dabchick on its nest, with chickweed adhering to its plumage.

The dabchick covering eggs with waterweed to camouflage before leaving the nest for a brief period.

The dabchick showing stained eggs, and its lobe-webbed feet.

As incubation usually commences before the clutch is complete, the chicks hatch at intervals. Here the first chick to hatch is shown sitting on its parent's back while the remaining eggs are incubated.

Walter Buller, writing in his *Manual of the Birds of New Zealand* (1882), remarks on the dabchick's quick reaction to threat:

> 'It dives with amazing agility, and, unless taken by surprise, will effectively dodge the gun by disappearing under the surface at the first flash, and before the charge of shot has reached it.'

New Zealand dabchicks have an extended nesting season. I have seen a nest with eggs in early July at Hamurana Springs, near Rotorua, concealed at the base of a sedge. In Northland I have seen nests with eggs as late as March. The late nesting may be a result of the loss of an earlier clutch, due to predation.

Most nests are no more than semi-floating masses of dead reed stems and waterweed anchored to stands of raupo or branches of willow trailing in the water. Some nests are built on solid ground among sedges at the water edge.

Top left: Small chicks ride on one parent's back while the other parent dives for food to feed them.

Bottom left: A chick reveals its lobe-webbed foot.

Above: A chick, two and a half weeks old, makes a vain attempt to ride on its parent's back.

On the Rotorua lakes dabchicks are known to build nests under the floors of lakeside boatsheds.

The usual clutch consists of two or three very pale blue eggs, and these soon become stained, as the birds cover the eggs with waterweed when leaving the nest for short periods. Both sexes share incubation for up to 22 days from the laying of the first egg, when incubation begins. At first, parents take short spells of incubation lasting about half an hour, but later will usually sit for over two hours. At this stage the sitting parent often appears reluctant to vacate the nest to allow its eager mate to take a turn at incubating. I have twice witnessed the incoming bird impatiently sitting on the back of its mate.

When approaching the nest, the dabchick dives several metres away and then surfaces close to the nest. Around some nests, chickweed floats on the water surface, and the birds' feathers become coated with a film of the weed.

Chicks hatch at intervals of up to two days, and the first chick to hatch mounts the parent's back as incubation of the remaining eggs continues.

When the last chick has hatched the nest is vacated and usually two chicks ride on one parent's back while it swims and the other parent dives for food to feed the mounted chicks. Before the swimming parent takes a turn at feeding, the bird raises itself, shaking the tiny chicks from its back. They then immediately scramble to mount the other parent. On two occasions I have seen one small chick remain on a parent's back and survive a reasonably long dive.

When chicks are a few days old they occasionally make short dives, but for several weeks they continue to be fed by their parents. Even when three or four weeks old the juveniles sometimes make futile attempts to mount a parent's back.

Chicks are usually independent when they are about 12 weeks of age. However, when early nesting occurs and a pair re-nests, younger chicks can feed themselves, being occasionally fed by one parent.

In late autumn most birds moult, congregating for safety on some of the larger lakes, particularly those in the southern North Island.

AUSTRALIAN LITTLE GREBE
Tachybaptus novaehollandiae novaehollandiae

Locally common • Protected
• Length: 25 cm • Status: Native

This Australian species was first recorded on Lake Hayes in Otago in 1968, and on a small lake near Dargaville in 1972 I had my first opportunity to photograph a pair.

The birds have now become well established on many North Island lakes, with a few sightings in the east of the South Island. They particularly favour the North Island dune lakes that are fringed with raupo.

These little grebes often share some habitats, such as relatively small farm lake locations, with New Zealand dabchicks, and the two species do not appear to compete for territory. However, the grebes are extremely wary when approached, quickly taking refuge in lake vegetation.

I visited a small raupo-fringed dune lake where an Australian little grebe had been recorded, and although I made a cautious approach, I could only see a pair of New Zealand dabchicks. It was not until I built a hide on the lakeshore that I was able to study the grebe at close quarters.

The Australian little grebe is minimally smaller than the New Zealand dabchick, and is easily distinguished by the prominent 'teardrop' of bare yellow skin below the eye and base of the bill. Its feeding habits appear to be similar to those of the New Zealand dabchick.

Like dabchicks, little grebes also build a semi-floating nest of dead reeds and waterweed anchored to surrounding water vegetation. The birds lay a clutch of up to four eggs, and from Australian records incubation is shared by the sexes for about 23 days.

An Australian little grebe.

Order Pelecaniformes

Shags

New Zealand has 14 species of shags, or cormorants. Most of these species inhabit marine environments, and only four species frequent freshwater habitats as well as coastal regions. These four species have black webbed feet in contrast to most marine shags with their pink or yellow feet, although a few marine species have black feet.

Shags feed on fish and crustaceans that they catch by diving from the water surface, creating scarcely a ripple. The exception is when little black shags feed in packs, taking it in turns to leapfrog over one another. Shags have strong hooked beaks that are ideal for gripping large fish or slippery eels. Fish are brought to the surface and juggled before being swallowed head-first. Eels are often a problem to subdue as they wriggle and coil around the shag's bill.

Shags use their webbed feet for propulsion when submerged, using their wings only when making sharp turns while pursuing their prey.

As shags' plumage is less waterproof than that of other water birds, such as ducks, they perch on rocks or posts after a spell of fishing with wings outspread to dry. I have seen them attempting to dry their wings even when in light rain. The lack of waterproof feathers may be an advantage as the completely waterproof plumage of other waterfowl makes them more buoyant, so it takes more energy to stay submerged when feeding.

When shags rise from the water, their wings flap heavily over the surface before the birds become airborne. They fly strongly in a direct flight with neck extended, only occasionally making short glides.

The birds nest in colonies, building substantial nests of sticks in trees close to water, sometimes on cliff ledges and, occasionally, as do black shags, on the ground on small secluded islets in lagoons.

Shags have variable and protracted nesting seasons, usually with peaks in spring and early summer and also in autumn. The usual clutches contain three or four chalk-encrusted white eggs that are incubated by both sexes in turn. The incubation period varies with species. Chicks are fed with regurgitated fish, with the chick often immersing its head deep inside the parent's open gullet.

BLACK SHAG
Phalacrocorax carbo novaehollandiae

- Cosmopolitan • Common; widespread
- Protected • Length: 88 cm
- Maori name: Kawau • Status: Native

Right: A black shag, showing the typical strong, hooked bill, common to all shags, ideal for gripping fish.

Black shags, also known as black cormorants, are a cosmopolitan species with many subspecies worldwide. The subspecies *novaehollandiae*, which is also found in Australia, is common and widespread in New Zealand, inhabiting sheltered coasts, rivers and freshwater lakes.

Like most other shags, they are extremely wary. In past years they were frequently shot due to their fish-eating habits, as fishermen thought the birds would compete for trout. Today they are protected. Outside the nesting season they are usually seen alone, and are easily distinguished by their large size and black plumage. But seen at

A pair of black shags at their nest.

close quarters the feathers reveal themselves to be bronze with black edging.

Black shags have a direct flight with rapid wing-beats and occasional glides, and are sometimes seen flying at considerable heights, especially when crossing land. They make a laboured rise from the water, flapping over the surface before becoming airborne.

The birds feed on a variety of fish that they catch by making a gentle dive from the surface, creating hardly a ripple. They remain submerged for up to 25 seconds, bringing their prey to the surface to be juggled and swallowed head-first.

Black shags have an extended nesting season, ranging from late winter to early summer, and peaking again in autumn. They nest in small colonies in trees near water, on cliff ledges of river gorges and lakes, and sometimes on the ground on small islets on lagoons.

Up to four eggs are laid at two-day intervals and are incubated for approximately 30 days by both sexes in turn. Chicks are fed in the usual shag manner on regurgitated fish, with the chicks often thrusting their head deep into the parent's gullet. The chicks leave the nest when about six or seven weeks old, but continue to be fed by their parents for up to three months.

PIED SHAG
Phalacrocorax varius varius

Locally common; widespread on sheltered coasts and inland lakes • Protected
• Length: 81 cm • Maori name: Karuhiruhi
• Status: Native

Pied shags, or pied cormorants as they are also known, inhabit freshwater lake and lagoon habitats, but they are most commonly seen in sheltered marine environments. The birds are distinctive with their long white throat and

A pied shag.

Top: A pied shag 'wing-drying'.

Above: A pied shag with a large flounder that it eventually managed to swallow whole.

Right: Chicks delve deeply into the shag's crop to receive their food.

underbelly and their black webbed feet.

The birds eat a variety of crustaceans and fish, including eel and flounder. Like other shags, pied shags sit low in the water, then with scarcely a splash they dive at intervals for fish, often remaining submerged for about 30 seconds. They then rest on the surface for 10 seconds before making further dives. The birds catch a variety of fish that are juggled on the surface before being swallowed head-first. In brackish lagoons pied shags often capture large flounder, which are dragged ashore to be subdued, but often prove difficult to swallow.

Compared with other shags, pied shags can be easily approached. Making a slow advance, I once even stroked an immature pied shag that was perched on a post.

Pied shags have an extended nesting season, especially when food is abundant. Nesting peaks in spring and early summer, and again in early

autumn. They nest in colonies, usually in trees, building a large platform of sticks and herbage. The birds will often pluck leaves and twigs from their host tree, and this, along with their caustic droppings, can denude a tree of its foliage.

They make a bulky nest of sticks lined with leaves, laying a clutch of up to five eggs at two-day intervals. Incubation lasts for approximately 30 days and is shared by both sexes. The chicks are fed regurgitated fish and usually feed by thrusting their heads deep inside the parent's gullet to grasp their meal. They leave the nest when about five weeks old, but are fed by the parents for up to three months.

LITTLE BLACK SHAG
Phalacrocorax sulcirostris

Common in North I.; widespread
- Protected • Length: 60 cm
- Maori name: Kawau-tui • Status: Native

Little black shags are common and widespread in wetlands throughout New Zealand, but are less common in the South Island. They are also found in Australia and New Caledonia.

After nesting, the birds feed in freshwater wetlands on a variety of small fish, insects and crustaceans. Little black shags often feed in packs of several birds. They herd small fish together and pursue the shoal by leapfrogging over one another, the birds in the rear of the pack leapfrogging to the front, and in turn being overtaken by birds from the rear. In addition to pack fishing in freshwater lakes and lagoons, they sometimes use the same method in sheltered marine habitats.

Little black shags can be distinguished from little shags by their shiny black plumage, long narrow bill and green eyes. Note that immature little shags are also totally black, but their

Below: Little black shags fishing in a pack in a marine habitat. They usually fish in lakes.

Right: Little black shags.

Little black shags nesting in trees beside a lake.

plumage is dull and the bill is short.

Little black shags have a more restricted nesting season than other shags, usually nesting in early summer, although a few occasionally nest in autumn within little shag colonies. They nest in small colonies, building a nest of sticks in trees overhanging water. In the Rotorua area some colonies nest on the ground on small islets.

Both sexes share incubation of the clutch of three to four eggs, with the eggs being laid at two-day intervals. Like other shags, the chicks are fed on regurgitated fish. At the time of writing, there appear to be no records of the incubation period or the time of fledging of the chicks, but they are possibly similar to those of the little shag.

LITTLE SHAG
Phalacrocorax melanoleucos brevirostris

Common; widespread • Protected • Length: 61 cm • Maori name: Kawau-paka • Status: Native

Little shags are common throughout New Zealand, inhabiting freshwater wetlands and sheltered coastal locations. A subspecies also inhabits Australia and New Caledonia.

The plumage of little shags varies from one bird to the next, some being pied with a completely white breast, others with a white throat, and some intermediate forms with a smudgy white and dark breast.

As with other shags, the diet consists of small fish, crustaceans and insects, which the birds

The white-throated plumage phase of the little shag.

catch by diving gently from the water surface. I have sometimes seen the birds feeding in very narrow farm drainage ditches where they catch small eels. Little shags feed in isolation and do not fish in packs as little black shags do. But I have seen several birds feeding together in shallow marine harbours.

Little shags have an extended nesting season starting in late winter and leading into summer, when they often share nesting trees with little black shags. They also nest among colonies of the large pied shag in pohutukawa trees in some sheltered marine locations.

Nests are built of sticks, and both sexes share incubation of the clutch of three or four eggs, which are laid at two-day intervals. As with other shags, chicks are fed on regurgitated fish.

Immature little shags have a completely black plumage until their first moult, and from a distance can be confused with little black shags, but the latter have a shiny black plumage, and the bill is longer and thinner than that of the little shag.

Above: Pied plumage of the little shag.

Top right: Little shags often fish on farm drains or very small ponds.

Bottom right: The 'smudgy' plumage phase of a little shag. White throated and pied plumage forms are more common.

Order Ciconiiformes

Herons

Six species of heron have been recorded in New Zealand, but only two species, the common white-faced heron and the white heron, are regularly seen in wetlands.

The reef heron, once our most common heron, inhabits only marine environments and does not visit inland freshwater wetlands, so is not included in this book.

The Australasian bittern that inhabits wetlands is classed in a subfamily of the Ciconiiformes, and the royal spoonbill is classed in a separate family.

Most of the species in this order are long-legged wading birds with large feet and un-webbed toes. They have long necks and long, sharp, pointed bills. Their broad wings are adapted so that the birds are able to fly distances with strong, slow wing-beats. Herons and bittern fly with their necks folded back, but spoonbills fly with their necks extended.

During the breeding season, some species sport long filamentous body plumes, while spoonbills display head crests. Although white-faced herons and bittern nest in isolation, white herons and spoonbills are colonial nesters.

All species feed on a variety of fish, crustaceans and insects, most being caught as the bird wades in shallow water, or stands motionless, waiting for prey to approach before capturing with a quick thrust of its sharp bill. Spoonbills are an exception. Although they catch fish, their broad, rounded (spatulate) bill is adapted to filter out small prey as they wade through shallow water, sweeping the bill from side to side.

After feeding on fish, particularly slimy eels, the bills of herons and bittern often become smeared with slime that cannot be removed with water. But the birds have adapted a method to remove it. They are equipped with patches of short, disintegrating, powdery feathers on their lower breasts and flanks, and when the soiled bill is rubbed into these feathers, a coating of fine talc adheres to it. The bird then rubs its bill on available foliage to complete the cleaning process and scratches the bill with a small, serrated comb on the end of the middle toe of the foot.

WHITE-FACED HERON
Ardea novaehollandiae novaehollandiae

Common; widespread • Protected • Length: 66 cm • Status: Native

White-faced herons are now our most common heron species; yet, prior to a large influx of this Australian species in the early 1950s, they were uncommon in New Zealand.

They have nested successfully and spread throughout the country, even being recorded in the Chatham Islands.

One reason for their success is that they feed in a wide range of habitats, including sheltered coasts and harbours, freshwater wetlands and at some times of the year in farm pastures, urban playing fields, and occasionally on the grass verges of roadways.

However, this chapter concentrates on the white-faced heron's habitation of freshwater wetlands, where they find an abundance of food, ranging from small fish to frogs, earthworms and insects. They feed by stalking in shallow water to catch their prey, or standing motionless as they wait for fish to approach before making

Right: A white-faced heron.

Top: A white-faced heron fishing in a lake.

Right: A heron raking to disturb mud.

Above: A white-faced heron in flight.

Far right: A white-faced heron at its nest over 20 metres up in a pine tree.

A white-faced heron pair greeting at the nest, with back plumes raised.

lightning-swift thrusts of their long bill to catch the prey, then swallowing the fish head-first. In shallow water herons also rake through the mud with their feet to disturb invertebrates. Sometimes they are seen stalking along the margins of wetlands where they search the herbage for insects.

Although the following incident occurred in a marine situation, it is worth recording the most unusual fishing behaviour by a white-faced heron that I have witnessed. The heron was perched on a beam of Ohope wharf in the Ohiwa Harbour, where it was intently watching a school of yellow-eyed mullet. Suddenly, the bird splashed into the deep water, with its wings outspread, seized a fish and returned to the wharf to swallow it. After a few minutes, the heron repeated the manoeuvre, so I ran for my camera and was able to record one more dive before the heron flew off. This is extremely rare behaviour for a white-faced heron, which normally feeds in shallow waters.

As mentioned in the introduction to the Ciconiiformes, when the birds have caught eels, the bill becomes soiled with slime, which is removed with the use of preening plumage as described on page 48.

Because white-faced herons depend on wetland locations for feeding, they sometimes nest in trees overhanging or near water. More often the birds build their nests high in pine or macrocarpa trees some distance from water. This

Top left: The parent's bill is seized crosswise by the chick to stimulate regurgitation of food.

Middle left and bottom left: After feeding the chicks, the parent rubs its bill in a powder-down patch, which renders the slime less viscous. The heron then combs and scratches off the combined matter with the serrated tip of its middle toe.

is no disadvantage, as they are strong fliers. Their flight is leisurely and graceful, with a strong downbeat of their wide, rounded wings.

When flying, the birds hold their heads in a folded-back position, although often when flying short distances, as when carrying sticks to build their nests, the neck remains outstretched.

In the warmer North Island, white-faced herons often begin courtship and nest building in winter months. The male, with back feathers erected, approaches the female, bowing and uttering croaks and guttural sounds, and then offers a stick to begin nest building.

The nests are usually built on the outer limbs of large pine trees. Unfortunately, early winter nests are often destroyed by storms, but nests built later in the season are more successful. In addition to pine and macrocarpa, some nests are placed in either tall kanuka or kahikatea, and others in gum trees. Near the coast nests are often built in pohutukawa trees.

Nests are composed of large sticks and lined with smaller twigs in which three or four pale turquoise-coloured eggs are laid. Incubation, which takes 23 or 24 days, is shared by the sexes and starts after the laying of the first or second egg, so that chicks hatch at intervals and are disproportionate in size. One parent constantly broods them for the first five or six days while the other flies off to find food.

With the return of the foraging parent there is a greeting ceremony of bowing, bill clicking and raising of back feathers. The chicks immediately clamour to be fed, but when they are very young, the incoming parent first settles to brood them for about 10 minutes, presumably to partly digest the food in its crop. The parent then

As chicks grow, they wander to perch on branches near the nest.

stands over the chicks as they thrust at its bill, and the bird's neck thickens as rhythmic regurgitatory movements start. The chicks, in turn, grip the parent's bill crosswise to receive their meal of partly digested fish. Feeding of the young chicks is repeated about every half hour. The interval between feeds lengthens as the chicks grow, and when they are near fledging, they may be fed only at intervals of three or four hours. At this time the chicks are fed undigested whole fish.

After feeding, the parent walks to a nearby branch where it begins a lengthy session of cleaning its bill and preening, out of reach of the clamouring chicks.

When chicks become feathered they begin to wander from the nest and to venture along nearby branches, stretching their necks and vigorously exercising their wings. Sometimes the chicks lose their balance, hanging upside down on the branch, but with much vigorous flapping they usually regain their upright position.

Being a veterinarian, I was occasionally brought chicks that had fallen from their nests. They were easy to rear, but quickly became imprinted and tame. After having been released in the wild and learning to forage for their food, these reared birds sometimes returned to my garden to feed. I would call with a whistle and the birds would return from some distance away to receive a meal of earthworms or mince meat.

Depending on the availability of food, chicks leave the nest when about six weeks of age. The older chicks are taught to forage for food by one parent while the other continues to visit and feed the remaining chicks.

From my observations, it appears that the birds do not return to the nest, and I have not seen a nest reused, although the same tree is sometimes used the following season. Most nests previously used disintegrate during storms.

I have sometimes seen white-faced herons nesting as late as summer. This late nesting is probably due to the loss of an earlier nest by predation or storm.

WHITE HERON
Egretta alba modesta

Uncommon; widespread • Protected
• Length: 92 cm • Maori name: Kotuku
• Status: Native

White herons are cosmopolitan, with more than four subspecies inhabiting both tropical and temperate regions. The subspecies of white heron found in New Zealand also occurs throughout Australasia, with the New Zealand population numbering approximately 200 birds. Adult white herons were recorded nesting in trees in a colony on the banks of the sluggish Waitangiroto River in Westland before European settlement of New Zealand.

The white heron occupies wetlands and some sheltered brackish lagoons, usually in autumn and winter, dispersing to these locations following nesting in the colony near Okarito in Westland.

A reserve has been created in the Okarito region by the Department of Conservation, and a substantial hide has been built on the riverbank opposite the colony for visitors to view the nesting birds.

White herons feed on a variety of fish, frogs and crustaceans. Like the white-faced heron, the birds catch their prey by stealthily stalking through shallow water, or standing motionless

After nesting, the white heron's bill changes colour to yellow.

until fish come within reach, before making rapid thrusts with their dagger-like bill to grasp the fish. Small prey are swallowed immediately and larger fish are juggled before being swallowed head-first. Like other herons and bittern, the birds clean the soiled bill by rubbing it in powder-down patches on their lower breasts and flanks, particularly after catching eels.

In late August and early September, white herons congregate at the Okarito nesting site. In September, the bill changes colour from yellow to black and the birds develop long filamentous plumes that fan across their back. They flare these plumes during courtship, as well as snapping their bill, bowing and uttering croaks and groans. During this courtship ritual, males make circling flights around the female.

Nests are built with sticks in the crowns of tree ferns or in kamahi and kowhai trees bordering the banks of the Waitangiroto River, which flows through the kahikatea swampland of the reserve.

A clutch of four or five pale turquoise-coloured eggs is laid in late September to October. Incubation of 24 or 25 days' duration is shared by the sexes, who feed the chicks with regurgitated fish. Chicks fledge when about six weeks old but stay around the colony, where they learn to feed and continue to be fed occasionally by the parents.

After nesting, the birds may still retain some of their plumes as the bill begins to turn golden, then fades to yellow. After the nesting season, birds disperse throughout both North and South Islands, inhabiting freshwater wetlands and brackish coastal lagoons.

Top left: A white heron.

Bottom left: The only white heron nesting colony is in trees on the banks of the Waitangiroto River in Westland.

Above: A pair of white herons at their nest.

Above: The vertically striped plumage provides good camouflage among brown rushes and weeds.

Above middle: Even when raupo swamps are green, a motionless bittern is hardly noticeable.

Above right: This bittern stood motionless for several minutes, with its bill close to the water surface, ready to grasp available prey.

Right: A bittern stalks slowly, searching for fish and frogs.

AUSTRALASIAN BITTERN
Botaurus poiciloptilus

Uncommon; widespread in wetlands • Protected • Length: 71 cm • Maori name: Matuku-hurepo • Status: Native

Bittern were once common in raupo swamps and freshwater wetlands enclosed with dense vegetation, but they have become less common, due to the drainage of many wetlands for agricultural purposes. They have also been affected by the loss of eggs and young to introduced mammalian predators, which have become more of a threat with the clearance of a great deal of wetland.

At the time when birds were shot for use as museum specimens, Walter Buller recorded how

common bittern were. In his *Manual of the Birds of New Zealand* (1882), he points to their abundance on swampy flats '. . . *on the west coast of the Wellington Province, where I have obtained half-a-dozen in the course of a single afternoon* . . .'.

Some current publications record that Australasian bittern are rare. I consider they should more accurately be described as uncommon, because the birds are often not sighted when they are standing in protective swamp foliage, particularly in beds of raupo. I have often carefully scoured an area of raupo where I knew birds were present, before locating one motionless bird, camouflaged effectively by its surroundings.

As they skulk through thick raupo stands, particularly when approaching their nests,

Left: A bittern at its nest in a raupo swamp.

Below: Before going on the nest to feed the chicks (left), the bittern regurgitates freshly caught prey, and (right) the chicks are then fed on partially digested fish or frogs.

bittern sometimes grasp several stems together with each foot. This provides a firm tripod-like perch on which they pause to cautiously stretch their neck above the rushes and scan the surroundings. They then lower the head to move on, and before long repeat the procedure some distance away, having obviously moved rapidly. Observing such furtive yet gentle behaviour one might wonder whether it was actually the head of a bittern that had appeared, or rather the leaves of raupo touched by a breeze.

Bittern feed on a wide range of fish, in particular eels. Frogs, freshwater crayfish and insects are also readily taken.

They tend to be crepuscular, feeding particularly late in the day or early in the morning, but when nesting bittern frequently feed during the day. The birds usually forage in shallow water along the edge of thick vegetation, but in the early morning will venture to feed in more open areas of swamp. The birds adopt a feeding method in which they slowly stretch their necks and hold their bills horizontally. In

After catching a slimy eel, this bittern's head is grey from cleaning with powder down.

this position both eyes are used to watch for prey, employing a binocular vision technique. When reaching a stretch of open water, they lower their head and stand motionless, with the head almost touching the water surface. I have watched them standing in this position for more than 10 minutes, before suddenly thrusting the bill forward to catch a fish.

During the day, I have often observed a bittern preening, with the bird partly hidden in foliage. Bittern preening can be a lengthy process and I have waited in my hide for two or three hours before the bird finishes and begins to feed.

In flight, bittern, like herons, fly with their necks folded and heads retracted. But bittern can be recognised by their broad, more rounded wings and brown mottled colouring.

In spring, male bittern produce mating and territorial calls that are low, resonant booming sounds. These have often been described as being similar to the sound of a bull, or the noise made by blowing across the neck of a large empty bottle. To me, the call very much resembles that of a male kakapo booming.

I once built a hide in a dense stand of raupo overlooking a small expanse of open water, hoping to see a bittern feeding. This did not eventuate, although I was able to film a harrier hovering only a metre above the water surface, and repeatedly dropping to catch some small prey, possibly tadpoles.

Later in the day, I heard a bittern boom close to the side of my hide. I watched through a peephole and was amazed to see the bird standing less than 2 metres away. And yet, even at such close quarters, the boom sounded surprisingly subdued. After about half a minute he repeated the call. The neck was stretched forward, then, following a long-drawn-out inspiration of air, he raised his head to make another resonant boom. He called three times before moving into thick cover. I was unable to record the event on film, as my camera tripod offered viewing from only the front of the hide.

After mating, male bittern take no part in sharing incubation or feeding the chicks. The female builds a well-hidden nest with dead raupo, reeds or rushes about 20 centimetres above the water surface in dense cover.

Some nests are occasionally built on solid ground. I discovered one in a stand of oioi (jointed rush), in a salt-marsh bordering a small river. I found the nest in this unusual position when studying fernbirds in the location. I noticed some very large footprints in a patch of mud and at first thought that they were those of a heron, but following them I discovered the nest with one bittern chick and an unhatched egg.

When the Atiamuri hydro dam was under construction on the upper Waikato River, before Lake Whakamaru had been formed, the area north of the river was covered with dense stands of raupo, stunted manuka and other vegetation.

A keen British birdwatcher was employed on the dam construction. He had watched bittern in the Norfolk Broads in England and in New Zealand he had located four nests in this area near Atiamuri. He contacted me for information about watching their nesting activities from hides, as he knew how extremely wary they were, particularly when nesting.

He eventually built camouflaged hides of hessian overlooking two of the nests, and when I left before dawn on a rare weekend off duty from my Warkworth veterinary practice, I was able to use his hides to study and photograph their nesting behaviour.

On one occasion, when the sun was shining behind the hide I was in, as the hessian-covered hide was rather flimsy, the nesting bittern had obviously seen my silhouette as I moved. She immediately stretched her neck vertically with bill pointing upwards so that she could watch me with binocular vision. I dared not move. She then slowly relaxed to continue brooding her two chicks and two unhatched eggs.

Bittern begin nesting from September to November, laying a clutch of up to five eggs. There appears to be some variation in intervals between egg laying. They are usually laid at two-day intervals, but I have noted that some eggs have been laid at daily intervals. Incubation starts after the laying of the first or second egg, and takes 23 to 25 days for each egg, so that chicks when hatched are of disproportionate size.

As mentioned earlier, the female alone feeds the chicks, so when she leaves the nest to feed, the chicks are vulnerable to predation by harriers. When returning to the nest she calls with a bubbling sound and very cautiously slides through the raupo stems, then carefully edges her large feet clear of the chicks.

When the chicks are very young they are unable to swallow large fish, so usually just before the bittern approaches the nest, she regurgitates any large, still-undigested fish or frogs at the back of the nest before feeding the chicks on her partly digested fish. She first feeds the clamouring oldest chick on the partly digested fish and, progressively, the younger chicks receive the more digested items.

After feeding the chicks, the parent swallows the remaining large fish deposited at the back of the nest.

As the chicks develop, in typical heron fashion they grip the parent's bill crosswise to stimulate regurgitation, but often items are spilt on the nest floor, to be quickly retrieved by the chicks. When the older chicks are two or three weeks of age they tend to wander from the nest where they are first fed by the parent. She then moves back to the nest to feed the younger chicks.

Small eels are often fed whole but partly digested to older chicks. And, like herons, the bittern uses the powder down on her flanks to clean off the eel slime, using the serrated middle toe to comb off the powder-covered slime. Sometimes the head of the bird is coloured grey with powder down. This may also appear on herons, but is less evident on their grey plumage.

I have been unable to discover how long fledged chicks are fed and when they cease to become dependent on the parent. There is still much to be learned about the behaviour of the Australasian bittern.

ROYAL SPOONBILL
Platalea regia

Cosmopolitan • Locally common; increasing • Protected • Length: 77 cm • Maori name: Kotuku-ngutupapa • Status: Native

The royal spoonbill, although classed in the Ciconiiformes order, belongs to the family that includes ibis and storks. Royal spoonbills differ from herons and bittern in a number of ways. These differences can be observed in flight, behaviour and feeding of chicks.

Left: A royal spoonbill.

Above: When royal spoonbills change over at the nest, after displaying, they often indulge in mutual preening.

Self-introduced to New Zealand from Australia, royal spoonbills first bred here in 1949, nesting in tall kahikatea trees close to the Okarito white heron colony. More recently, some birds from this small colony, possibly reinforced by the arrival of more birds from Australia, formed nests in a larger colony on the ground, on small sandy islands in the Vernon Lagoons in Marlborough. It was here that I studied and

Top: Royal spoonbill parents take turns at incubation. At the changeover of feeding or incubation the parents perform a greeting ceremony, clicking bills and erecting crest feathers.

Left: The method of feeding the chicks is similar to that of shags and ibis (unlike herons), with the chick inserting its bill into the parent's bill to receive food.

Above: A newly hatched royal spoonbill chick.

Spoonbills fly with necks extended.

photographed them in 1978. They bred successfully and later established new colonies on Kapiti Island and several colonies on the east coast of Otago. More recently, spoonbills established a colony in the Parengarenga Harbour in Northland and are increasing in number there.

The birds feed on a variety of fish, which are caught by stalking in shallow swamps. However, their more usual method is to feed in groups, slowly wading through shallow water and sweeping the spatulate bill from side to side as they sieve small aquatic organisms through its serrated edges. As the bill is sensitive, the birds do not need to rely on sight and are able to fish in muddy water, as well as at night.

Royal spoonbills begin nesting in late September, building substantial nests of sticks and seaweed lined with grasses. Parents share incubation of the three or four eggs for 25 days, and when changing over at the nest they clatter their bills and erect their head crests in greeting. They also briefly preen each other's necks.

Unlike young herons, spoonbill chicks do not grip the parent's bill crosswise, but thrust their short bill into the parent's open bill to grasp the regurgitated food. Chicks leave the nest when about four weeks old, but they stay with their parents to feed for several weeks.

Immature birds can be recognised in flight by the black edging of their wings, and, unlike herons, royal spoonbills fly with necks outstretched, often circling in small groups.

After the nesting season, spoonbills disperse throughout New Zealand, but they are seen more frequently on North Island coastal lagoons and harbours. Populations of spoonbills are also regularly seen on the Manawatu Estuary and the recently dismantled sewage oxidation ponds now open to the Manukau Harbour.

Order Anseriformes

Swans, Ducks, Geese

Most members of the order are fully web-footed and strong fliers. The exceptions are the Auckland Island and Campbell Island teals, which are extremely rare and flightless. They survive in marine habitats and are not found in freshwater wetlands.

Of the 11 species that inhabit freshwater wetlands in New Zealand, five are endemic, two are native and four species are introduced.

The largest of these birds is the mute swan, which was introduced to New Zealand from Britain in the 1860s. They were introduced for ornamental and sentimental reasons and are not common. Although they inhabit some urban parks, the main population survives on Lake Ellesmere.

The very common, smaller black swan was introduced from Australia, mainly for game purposes in the 1880s, but as the population increased remarkably, it is probable that large numbers introduced themselves to New Zealand in the following years.

In the 1870s Canada geese were introduced, but they did not become common until a further introduction was made in the 1900s.

The commonest and most widespread of the dabbling ducks is the introduced mallard, which, due to its aggressive habits, often interbreeds with the native grey duck.

The endemic paradise shelduck is common in some South Island locations with numbers now increasing throughout the North Island.

The threatened endemic blue duck is widespread in subalpine rivers, from the central North Island to Fiordland.

The native grey teal is widely distributed throughout freshwater wetlands in New Zealand. Being a strong flier it often commutes to and from Australia.

The endemic brown teal, once common, particularly in Northland, is now an endangered species, with the strongest populations inhabiting the wetlands of Great Barrier Island and Fiordland.

The endemic New Zealand shoveler, the fastest flying duck, is widely distributed in shallow wetlands throughout New Zealand.

Our only true diving duck, the endemic New Zealand scaup or black teal, is widespread in clear, deep-water lakes throughout the country. It also inhabits some hydroelectric lakes.

MUTE SWAN
Cygnus olor

Uncommon • Protected • Length: 150 cm • Status: Introduced

The beautiful mute swan, or white swan, was introduced from Britain in the 1860s for ornamental and sentimental reasons. It now inhabits a few urban park lakes and some wetland lakes, but the main population is resident on Lake Ellesmere. Despite their common name, the birds are often quite vocal.

Mute swans feed on aquatic plants and graze on grasses and vegetation along the verges of lakes. It has also been reported that they take aquatic insects.

In Britain, mute swans nest in large colonies and pair for life. In New Zealand, they usually nest in isolation, building a huge mound of a nest composed of leaves of raupo, flax leaves and

Mute swans.

other vegetation. The nest is lined with a few feathers and down.

The nesting season begins in early summer, with a clutch of up to eight white eggs being laid. The female (pen) incubates them for about 35 days, while the male (cob) aggressively defends the nest.

The chicks (cygnets) are able to swim and feed soon after hatching, although they remain with the parents for several weeks.

BLACK SWAN
Cygnus atratus

Common; widespread • Protected, except in shooting seeason • Length: 120 cm • Status: Introduced

Black swans were first introduced to New Zealand from Australia during the 1860s. But since the population increased markedly a few years later, it is probable that large numbers migrated to New Zealand following a drought period in Australia.

The birds now occupy freshwater wetlands throughout New Zealand, and also inhabit many marine harbours.

Black swans fly strongly with neck extended, and when in flight the white ends of their wings are evident. When taking flight from water they flap along the surface before becoming airborne.

Their food consists of aquatic plants, which they often have to reach from the bottom of a lake by upending. They also graze farm pastures close to lakes, making the herbage unpalatable for stock to graze.

Black swans have a very extended nesting season. In northern regions I have found single pairs nesting in midwinter. They also nest as late as January.

Their nests are bulky structures, composed of raupo and rushes and other vegetation, and often built close to the edge of a lake. But in Lake Ellesmere, which is home to very large numbers of the birds, the nest is often built on dry land some distance from the water.

The usual clutch consists of up to six very pale green eggs that are incubated in turn by the pair for 35 to 40 days. After hatching, the chicks (cygnets) are active and able to swim within a couple of days. Both parents care for the chicks, with the male being particularly aggressive towards any other approaching bird.

In autumn the birds congregate to moult in secluded parts of lakes and particularly in sheltered marine harbours.

Above: Black swans with cygnets.

Far left: A black swan and its nest.

Left: Black swans show the white of their wings when in flight.

CANADA GOOSE
Branta canadensis maxima

> Common; widespread on grasslands near waterways • Length: 83 cm
> • Status: Introduced

A liberation of a small number of Canada geese from North America was made in the 1870s. But it was not until the early 1900s, after a larger release, that the birds became established. The birds are now very common in the eastern South Island and in the tussock country of eastern Fiordland. Large populations inhabit Lake Ellesmere, and introductions to the North Island have resulted in successful breeding.

Canada geese are strong fliers, and when several birds fly together for any distance they adopt a 'V' formation.

Canada geese feed on a variety of herbage, often causing considerable damage to pastures and sometimes to standing crops. But during the

Left: Canada geese with their goslings. Above: A Canada goose incubating its eggs.

nesting season, the birds tend to favour freshwater wetlands and lakes where they feed mainly on aquatic plants and graze the edges of lakes with their young.

Nesting begins in early spring and extends to early summer. In the South Island they commonly nest in the upper reaches of high-country rivers, and other nesting sites are located close to water. Nests consist of rushes and dried grasses and are lined with down. The usual clutch consists of four to eight white eggs that are incubated by the female while the male guards the nesting area.

Goslings hatch after 28 days of incubation and are able to swim soon after hatching, though they remain with the parents. However, when several birds nest in close proximity, crèches are formed, with the adults taking turns to guard.

PARADISE SHELDUCK
Tadorna variegata

> Common; widespread in open areas
> • Length: 63 cm • Maori name: Putangitangi • Status: Endemic

Unlike the European shelduck of the same genus, in which the male and female are almost indistinguishable in colour, the male and female paradise shelduck differ markedly in plumage colouration. The male sports an overall dark plumage with dark head, while the female has a conspicuous white head and upper neck and a more highly coloured body plumage. Paradise shelducks are larger bodied than ducks and show certain gooselike characteristics.

The birds are widely distributed throughout New Zealand wetlands. They are also seen in pairs on sheltered coastal beaches, in alpine regions and on open farmlands.

They pair for life, and during the nesting season often inhabit farm pastures close to wetlands, or farm pastures where the only water may be a small farm pond.

Paradise shelducks fly with strong wing-beats, which are considerably slower than those of ducks, and when flying in flocks for any distance adopt a 'V' formation. Before moulting in late summer, the birds congregate on secluded lakes and lagoons. After moulting, pairs claim their nesting territory.

Their food consists of grasses, seeds, insects, earthworms and other invertebrates, and aquatic plants in wetlands. They also occasionally feed in sheltered harbours and marine estuaries.

In northern districts birds begin nesting in early August, and later in southern regions. They usually choose to nest in burrows of a bank, under a fallen log or beneath tree roots. In the north, a favourite location may be a hole in a mature puriri tree.

Small collections of dried grasses and leaves are used for the nest, which is lined with down, and a clutch of six to nine eggs is incubated by the

female only, while the male guards the location.

Ducklings hatch after 30 to 35 days, and soon after hatching they accompany the parents to the nearest water, which may just be a small farm pond. When approached by an intruder the parents attempt to attract attention away from the area with much loud discordant honking and guttural calls. The swimming ducklings are able to dive below the water surface to elude intruders, even when small.

After about two months the ducklings are feathered in dark plumage, somewhat resembling that of the male, but females display a white ring around the eyes. They do not assume adult plumage until after the first moult.

Top: Paradise shelducks with ducklings. Soon after hatching, ducklings accompany the parents to the nearest water. Above: A female paradise shelduck with its ducklings.

BLUE DUCK
Hymenolaimus malacorhynchos

Uncommon; widespread • Length: 53 cm • Maori name: Whio • Status: Endemic

The endemic blue duck inhabits turbulent alpine and subalpine rivers and streams, particularly those fringed with trees or thick herbage, or steep-sided gorges. Blue ducks are not found north of the central North Island. They appear to remain in particular river catchments and, except when young, do not move to other stretches of river.

They are usually seen in pairs, and when standing motionless on river rocks their blue-grey colouration blends exactly with the rocks or boulders, and the birds are often noticed only when one moves its head, revealing the light-coloured bill.

Blue ducks are strong fliers, flying low, directly

Above: A blue duck and its ducklings rest on a rock.

Right: Blue ducks are at home in a turbulent mountain stream habitat.

Left: Blue ducks. Their bills have fleshy edges, which protect them from injury while plucking prey off underwater rocks. Above: Blue ducks with ducklings, Whirinaki River. Ducklings stay with the parents until fully feathered.

over the surface of the water to another stretch of river when disturbed. They possess a strong homing instinct, which was demonstrated when conservation work was undertaken to relocate identified banded birds to another region many kilometres away, and the birds were later found to have returned to their home territory.

Blue ducks feed on a variety of aquatic insects, particularly caddis fly larvae and insects that have fallen from surrounding trees. Their bills have fleshy edges, which likely serve to protect them from trauma when the birds pluck prey adhering to underwater rocks.

Blue ducks begin nesting in early spring, but also as late as summer. However, the later nesting may be due to the loss of an earlier nest by predation. Nests, composed of grasses and a few sticks and lined with down, are well hidden in dense vegetation, rock crevices or under fallen logs near riverbanks. The usual clutch is seven or eight eggs, which are incubated by the female for about five weeks.

Ducklings are able to swim and dive when a few days old, but stay with the parents until fully feathered at about three months of age. They then move away from their natal territory, although they usually remain in the same river catchment.

MALLARD
Anas platyrhynchos platyrhynchos

Abundant; widespread • Length: 58 cm
• Status: Introduced

Since their introduction for game purposes mallard have become the most numerous and widespread of waterfowl. They inhabit all types of wetlands including urban parks and, particularly during the shooting season, large flocks frequent sheltered marine habitats.

From British stock in Australia, mallard were first introduced to New Zealand in 1867, with later introductions being made from America and Britain.

Mallard feed on aquatic vegetation and invertebrates, particularly snails, finding these along the water's edge, and in shallow rivers and waterways. They also graze across grassland, taking seeds and insects.

The birds are able to breed when a year old and, being aggressive, they also interbreed with the native grey duck. Males in breeding plumage are brightly coloured but females are drab, being

brown flecked, and it is possible to confuse them with grey ducks. However, they can be distinguished from grey ducks by their orange legs and purple speculum, whereas grey ducks have a green speculum.

Mallard begin nesting in winter months and lay clutches of eight or more eggs in nests of grasses lined with down. Nests are built in thick herbage, usually close to water. The female incubates for approximately 28 days and the ducklings are taken to water soon after hatching.

Left: A mallard showing the blue wing speculum.

Above: A mallard pair: the colourful male with female.

They fledge when about eight weeks old.

Like all ducks, mallard congregate in flocks on lakes to moult in late November or December, when they become flightless for about three weeks. During this period they stay away from the shore and only land to feed at night.

GREY DUCK
Anas superciliosa superciliosa

> Common • Partially protected
> • Length: 55 cm • Maori name: Parera
> • Status: Native

Grey duck also inhabit Australia. In New Zealand they prefer slow-moving rivers, small lakes and freshwater wetlands where there is open water and beds of rushes. The birds do not normally inhabit urban park lakes and are less confiding than mallards.

Sexes are similar in plumage colour, with the birds being darker than female mallards. They have a prominent dark eye stripe and dark coloured legs and can also be distinguished from female mallard by their green speculum. This feature is noticeable when they have interbred with mallard, and their legs are often a yellowish colour like those of the mallard.

Grey duck feed on aquatic vegetation, seeds and invertebrates, especially earthworms and snails.

The birds are very early nesters. In northern regions I have seen ducks with ducklings in June. Nests built with grasses and lined with down are hidden in thick herbage and sometimes in the holes or forks of trees.

For three consecutive years a grey duck nested in the cupped fork of a tree, no less than 8 metres above the ground in Kowhai Park, Warkworth. The fluffy, lightweight ducklings were uninjured when, after hatching, they fell to the ground and followed their parents to the nearby Mahurangi River.

A grey duck showing green wing speculum.

GREY TEAL
Anas gracilis

- Locally common; widespread • Protected
- Length: 43 cm • Maori name: Tete
- Status: Native

Grey teal were uncommon in New Zealand up to the 1970s, but after that time large numbers migrated from Australia. Following provision of artificial nest boxes in many freshwater wetlands, they have increased remarkably in number and are now widespread throughout New Zealand.

The birds are one of our smaller dabbling ducks. Although males and females are similar in colour to females of other species, grey teal lack facial markings.

Grey teal are now common in many lagoons and raupo-fringed wetlands, where they strip seeds from overhanging plants. They also feed on aquatic plants, invertebrates and small organisms. They obtain the last by dabbling in shallow water and ooze to filter out the small food items, often moving in unison.

It is thought that grey teal pair for life; nonetheless, like other ducks they congregate in large flocks during the moulting season. The birds are agile in flight, and able to fly straight upwards when disturbed.

A pair of grey teal dabbling in ooze.

A grey teal with duckling.

Although the birds are known to nest in early summer months, they usually start nesting in winter months. The nest is constructed of a bowl of dried grasses lined with down and built in thick vegetation close to the water's edge. Clutches generally consist of up to eight eggs, which are incubated by the female teal for 27 or 28 days.

The male stays near the nest during this period. Ducklings take to the water within a few hours of hatching, becoming independent and dispersing when about eight weeks old.

BROWN TEAL
Anas aucklandica chlorotis

Endangered ● Protected
● Length: 48 cm ● Maori name: Pateke
● Status: Endemic

Brown teal were once found throughout many wetlands in both the North and South Islands. But due mainly to predation, they are now restricted to Northland and Fiordland, with the main population occupying wetlands within

Great Barrier Island. The birds may also be found on tidal streams and in sheltered marine harbours of the island. However, even on this island, which is free from possums and mustelids, feral cats and dogs prey on the birds.

In the late 1950s brown teal were present on tributaries of the Waipu River, and it was here that the late Sir Peter Scott was taken to view them. At this time the crepuscular brown teal used to roost and nest in thick vegetation on the banks of these streams. But due partly to modern farming practices, the increased grazing and clearance of vegetation resulted in a loss of sheltered roosting sites for the birds, and gave them no protection from predators.

Conservationists found that brown teal adapted well to captivity, with the birds then able

A brown teal pair. The main population of brown teal occurs in wetlands on Great Barrier Island.

to be introduced to predator-free islands. In addition, captive-bred birds have recently been introduced to predator-free sanctuaries on the mainland. And, due to control of predators, brown teal are now becoming established in many Northland wetland habitats.

Brown teal feed on aquatic plants and invertebrates, and dabble in shallow water to filter out small food items.

The birds nest in dense vegetation, often some distance from water. A clutch of five or six eggs is incubated by the female for 27 or 28 days, with the chicks fledging and becoming independent when eight weeks old.

NEW ZEALAND SHOVELER
Anas rhynchotis variegata

> Widespread • Partially protected
> • Length: 48 cm • Maori Name:
> Kuruwhengi • Status: Endemic

New Zealand shovelers are our fastest flying ducks. They rise immediately from the water surface and can be recognised by their prominent white wing-bars and rapid flight.

The birds inhabit fertile lowland swamps, lakes and lagoons that are bordered with raupo, and they are uncommon on flowing rivers and high-country lakes.

Shovelers sit low on the water and can be identified by the large wedge-shaped bill, which is modified with prominent lamellae along its edges for filtering small food items from the water. The birds also eat aquatic vegetation and larger invertebrates, including water beetles and freshwater snails.

Male shovelers display a colourful breeding plumage, and even in eclipse plumage can be identified by the white crescent at the base of the bill. The female's plumage is drab in comparison.

Like other ducks, they congregate to moult after the nesting season, and often pair off and start courtship when still in winter flocks, yet do not nest until late August and September.

Nests are well hidden in thick vegetation, often some distance from water. The female incubates the eight or more pale blue eggs for about 27 or 28 days. The male usually stays guarding the nest and joins the female when she leaves the nest to feed.

Unlike other dabbling ducks, shoveler ducklings stay close to thick cover at the water's edge to feed, leaving their parents and dispersing when about eight or nine weeks old.

Below: A male New Zealand shoveler, and (bottom) a female New Zealand shoveler.

NEW ZEALAND SCAUP
Aythya novaeseelandiae

> Uncommon; widespread • Protected
> • Length: 40 cm • Maori name: Papango
> • Status: Endemic

New Zealand scaup, or black teal as they are also known, are our smallest ducks as well as our only true diving duck. The birds are easily distinguishable by their dark, glossy plumage. Males can be recognised by their golden-coloured eyes; the female's eyes are brown.

They inhabit clear, deep freshwater lakes, including many hydro lakes and dune lakes. The birds are also found on some alpine tarns. Scaup do not normally inhabit shallow wetland swamps or flowing water. The birds are almost entirely aquatic and are infrequently seen on land.

With very efficient diving skills, they dive from the surface to feed quite deeply on aquatic

plants and invertebrates, usually remaining submerged for about 15 seconds, but often for as long as 40 seconds. The birds also feed on surface insects and algae.

Scaup fly rapidly, revealing their white underwings. In autumn they congregate in large flocks, with some flocks consisting only of males.

The birds then pair off in early winter. They nest in protective vegetation close to the water, with the female building a bowl-shaped nest of dried grasses thickly lined with down. The clutch of six to 10 large eggs is surprising for this small duck. The male stays close to the nest while the female incubates for about 28 days. The ducklings are able to swim and dive when a few days old, but remain with the parents for up to eight weeks before dispersing.

Top: The male New Zealand scaup has a bright golden-coloured eye.

Above: A female New Zealand scaup with ducklings.

Order Falconiformes

Diurnal birds of prey

The only bird of prey found inhabiting New Zealand wetlands is the Australasian harrier, which is widespread and common. This species is also common in Australia and many Pacific islands.

Unlike the falcons in this order, which have dark eyes, harriers have a yellow or orange-coloured iris and lack the tubercle in the nostril.

Their feeding habits differ too, with harriers regularly feeding on carrion as well as catching live prey.

In birds of this order, the females are markedly larger than the males.

AUSTRALASIAN HARRIER
Circus approximans

Common; widespread in open country
- Protected • Length: 55–60 cm
- Maori name: Kahu • Status: Native

Australasian harriers inhabit coastal areas, forest margins, alpine tussock and open farmland. This chapter deals with the birds' reliance on wetlands for much of their food and for nesting sites. Harriers also sometimes gather at communal roosts in swamps at dusk.

It is possible to determine the age of harriers approximately, as in juveniles the plumage is a deep brown and, with age, this fades to a generalised buff colour.

Harriers are often noticed as they quarter the ground searching stands of raupo and swamp vegetation where they take as prey fernbirds, nesting banded rails and crakes. They also catch frogs and are reported to sometimes take fish in shallow water. On page 62 I refer to my

Above: A harrier feeding on carrion.

Right: An Australasian harrier.

Left: A harrier at its nest in a swamp.

Above: Australasian harrier chicks.

observation of a harrier hovering low over shallows in which I was studying bittern. On another occasion, from a viewpoint on a hill overlooking a lagoon, I watched a harrier circling over dense raupo, before it suddenly dived on a duck with small ducklings. There was much splashing and then all was quiet. I watched for several minutes expecting that the bird had captured a duckling, before deciding to investigate. The harrier flew off and I found it had caught the parent grey duck, leaving it partially plucked, and the ducklings had disappeared into the raupo.

Pairing begins in winter, when birds soar high in the air, performing courtship dives and calling with a repeated, high-pitched 'kee-kee' whistle. They also perform aerial courtship displays that tend to be above, or near, the nest sites.

The birds usually begin to nest in early August in northern districts. Nests are bulky structures of sticks, raupo leaves and grasses built among dense raupo stands or near the edges of swamps in dense vegetation.

A clutch of four or five eggs is laid and incubation begins after laying of the first egg. This results in eggs hatching at two-day intervals, and the last chicks to hatch seldom survive. The female feeds the chicks with food sometimes brought to the nest by the male, but usually food is passed to the female in an aerial transfer.

Early naturalists reported that harriers often deserted their nests when they were discovered. In one instance, I was walking along the edge of a raupo swamp and in a small clearing I caught sight of a harrier on her nest. I immediately averted my gaze and continued walking away from the swamp. However, when I returned to the site five days later, I found the bird had deserted her clutch of four eggs.

The chicks usually fledge when six weeks old, continuing to be fed by the parents for about one more week.

Order Gruiformes

Rails, Swamphens and Coots

Many species in this order inhabit freshwater wetlands. All members of the order have rounded wings, and although infrequently seen flying during the day, they often migrate to other wetlands at night.

The rails and crakes have laterally compressed bodies that enable them to move quickly through rushes and thick vegetation. To quote Walter Buller's *Manual of the Birds of New Zealand* when describing a crake: '*Its compressed form enables it to thread its way among the close-growing reed stems with wonderful celerity.*'

All species except the coot have unwebbed feet, yet are good swimmers. The coot, which is the most aquatic of the family, has lobe-webbed toes, somewhat similar to those of the grebes. The coot and the pukeko sport bony frontal shields.

With the exception of the pukeko, most rails that inhabit wetlands are secretive and tend to remain in the cover of thick vegetation.

BANDED RAIL
Rallus philippensis assimilis

Locally common, in Northland, Great Barrier I. and northern South I.
- Protected • Length: 30 cm
- Maori name: Moho-pereru
- Status: Native

Above: A banded rail. Right: On the nest.

Banded rail were once widespread across New Zealand, but have suffered from a loss of wetland habitat as well as predation. They now occur mainly in mangrove swamps and salt-marshes, but also inhabit coastal freshwater wetlands and coastal farm drainage streams.

Subspecies of the banded rail inhabit Australia, Indonesia and many Pacific islands. In some of these islands the birds are extremely confiding, even entering motels and shopping areas. However, in New Zealand banded rail are usually wary and secretive, particularly when nesting. Yet on one occasion a friend was walking along the edge of the Clevedon River when he was twice pecked on the back of his leg. He turned to see a highly aggressive banded rail, with feathers puffed, which he later found had a nest nearby in thick herbage.

Banded rails tend to feed in the mornings and evenings, eating a mixed diet of aquatic insects, earthworms, small fish and tadpoles. They also take the seeds and fruits of aquatic plants. Where the birds inhabit coastal swampland, they feed on crabs, crustaceans and molluscs.

The birds have a very extended nesting season. In northern regions, I have found nests in July and

A banded rail pulling reeds to form a bower over its nest.

as late as March. These late nests are probably due to loss by predation of earlier nests.

Their favourite nesting locations appear to be in stands of oioi (jointed rush), but I have also found nests in thick wetland herbage and dense kikuyu grass. To hide their nests from aerial predators, herbage is pulled over to form a bower. I have found the birds to be extremely cautious when nesting, and very wary of the slightest unfamiliar sound. For example, they are alert to the purring of a movie camera motor.

The usual clutch consists of three to five eggs that are incubated for 22 to 24 days by each parent in turn. Chicks when first hatched are clothed in sooty-coloured down. They leave the nest soon after hatching and forage for food with their parents until they become independent at about eight weeks old.

A spotless crake coming to its nest.

SPOTLESS CRAKE
Porzana tabuensis plumbea

Locally common; widespread
- Length: 20 cm • Maori name: Puweto
- Status: Native

The diminutive spotless crake is a shy, secretive occupant of the wetlands, particularly the shallow sedge and raupo swamps of the North Island. The birds are less common in the South Island, where the marsh crakes, although similarly cautious, are more common and occupy a similar territory.

An exception to the wary spotless crake of the mainland wetlands is the spotless crake that is found on the Poor Knights Islands. Because of the open, low-lying shrubby vegetation of this island group and a lack of predators, the birds have no need for concealment and instead may be seen foraging in the open and nesting on the bare ground.

The spotless crake is also common in Australia and many of the Pacific islands.

Due to their secretive habits, little is known of the birds' behaviour and consequently they are often considered uncommon. However, spotless

Left: A spotless crake on its nest.

Above: A spotless crake. This shows its small size.

crakes appear to be quite widespread, even in small wetland locations. This has been proved when tape recordings of their calls have been played. The birds will readily reply with rattling calls followed by repeated 'pip-pip' sounds. But even with this contact, they do not usually leave the cover of vegetation or come into view.

These small (blackbird-sized) birds are most typically seen when individuals have been caught by domestic cats and brought to the house. The first time I saw a spotless crake was when one had been caught by a cat and brought to a farmhouse from a very small wetland site nearby.

Like other members of the rail family, spotless crakes possess a very strong homing instinct.

Some years ago, members of the Hamilton Junior Naturalists Club were concerned about a small raupo swamp that was to be drained for farming purposes. They arranged to net the birds and relocate them. The birds were leg-banded for identification and released in a similar wetland more than 25 kilometres away. However, three weeks later, when members revisited the original habitat to check that all the birds had been captured, they found that the same banded spotless crakes were being caught. It is thought that the birds fly at night when they migrate to other swamplands.

Crakes are generally crepuscular in habit, being most active at dusk and dawn. They feed on a large variety of insects, spiders, grubs, earthworms and tadpoles. They also take seeds from wetland plants and other wind-blown seeds from nearby grassland.

In both the North and South Islands, spotless crakes build their nests well above the waterline, often along the edges of lagoons or shallow lakes. The nests, formed of loosely woven sedge, strands of grass and dead raupo leaves, are often built in cutty grass among clumps of raupo or in sedges. Where there are willow trees, the birds favour the extra shade and concealment of the low trailing branches.

During the nesting season, nests are usually difficult to find, being well hidden among raupo and grasses. However, in winter, when the raupo stems have died and flattened, the clumps of bright green cutty grass are prominent and old nest sites are easily located.

Spotless crakes lay two to four lightly spotted eggs that are incubated by both parents in turn for almost three weeks. The newly hatched, sooty-coated chicks are precocial and leave the nest after two or three days. They then quickly learn to find and feed on insects and grubs, but remain with their parents for several weeks.

MARSH CRAKE
Porzana pusilla affinis

Uncommon in North I.; more common in South I. • Protected • Length: 19 cm • Maori name: Koitareke • Status: Native

Little is known of the habits of the secretive marsh crake, which is also found in Australia. Like the spotless crake, these birds are only the size of a blackbird. The nominate subspecies is also found in many wetland locations in the northern hemisphere. However, it is not appropriate to use the published behaviour of this subspecies for marsh crakes in New Zealand, where there is little published information.

On two occasions while I was rowing a boat on a narrow tributary of the Wairau River in the South Island, where the banks were covered in thick vegetation, marsh crakes were flushed out by the disturbance I made. The birds flew close to me with legs trailing and then suddenly landed in thick cover. Apart from a brief sighting in Lake Alexandrina, these are the only occasions I have seen a marsh crake.

It seems that the birds fly at night, when they

Marsh crake. (Focus NZ Photo Library)

have been reported away from their usual habitats. At dusk and night, marsh crakes make a variety of calls, mostly a harsh 'krek'.

The birds feed on a mixture of invertebrates as well as the seeds of aquatic plants. Rarely seen in the open, they prefer the cover of swamp vegetation, although the birds sometimes feed in the tidal margins of salt-marshes.

PUKEKO
Porphyrio porphyrio melanotus

Widespread; abundant • Protected
• Length: 51 cm • Status: Native

Subspecies of the New Zealand pukeko are common in many overseas countries where they are known as purple gallinules. In New Zealand pukeko are also known by the name swamphen or purple swamphen.

Unlike the secretive rails, the brightly coloured pukeko is a familiar bird seen in many wetland habitats throughout New Zealand. Having benefited from land clearance, the birds are frequently seen feeding on herbage beside busy roads and are common in many urban parks. Pukeko often live in groups and are usually approachable, having become used to the presence of humans.

However, they prefer swamps with ample vegetation. Their large feet support them on marshy ground, and although not web-footed, they swim well. While their flight may appear clumsy, pukeko can fly long distances.

Pukeko are omnivorous, taking vegetation, invertebrates, small fish and frogs. They also rob

A pukeko and a newly hatched chick.

chicks. In these cases a nest may contain 12 eggs. Incubation takes about 24 days and the chicks remain around the nest site for three or four days after hatching.

AUSTRALIAN COOT
Fulica atra australis

Cosmopolitan • Protected
• Length: 38 cm • Status: Native

Australian coots were first recorded in 1958 on Lake Hayes in the South Island. They have bred successfully and are now common on many clear-water wetlands and lakes that are preferably fringed with raupo and thick vegetation. They are also common in many urban parks where they mingle with ducks and swans.

Coots are the most aquatic of the rails and infrequently walk on land as they have lobe-webbed toes, somewhat similar to those of grebes. The birds are not often seen flying during the daytime, but they are strong fliers. However, they have a laboured flight, pattering along the surface before becoming airborne.

the eggs of some ground-nesting birds. When feeding on stems of vegetation, the food is held with one foot and is eaten parrot fashion.

The birds have a very extended nesting season. I have found nests in winter months and on one occasion in late February. Pairs will nest singly, building deep bowls of interwoven grasses that are well hidden in thick vegetation. But frequently a number of birds form a nesting community of two or three females laying eggs in one nest, with several members of an extended family taking turns at incubating and feeding the

An Australian coot and chicks.

Above: An Australian coot feeds its chick.

Right: Australian coots chasing a black swan from their territory.

The birds feed on aquatic plants, upending and bringing vegetation to the surface to sort out the succulent shoots. Although largely vegetarian, coots search out invertebrates among the waterweed, particularly water snails. The birds also forage on grass close by the water's edge, looking for grubs and insects, which are valuable for extra protein when there are chicks to feed.

Coots usually raise three broods a year, nesting on the shores of lakes in the North and South Islands. The nest consists of flattened raupo leaves and other vegetation and is built close to the water. Coots are very aggressive when nesting and will even attack black swans that venture near the nest.

Parents take turns incubating the clutch of four or five eggs for three weeks. Chicks stay near the nest after hatching and make brief sorties to be fed by the parents. In autumn, coots moult their feathers over a short time, gathering together on large lakes.

Order Charadriiformes

Waders, Gulls and Terns

A large number of wetland birds appear in this order. The species dealt with in the following descriptions feed and nest in a variety of wetland locations, including braided river wetlands. Some species migrate to marine habitats at certain times of the year.

These birds include the very common South Island pied oystercatcher, usually referred to as SIPO, the common pied stilt, the extremely endangered black stilt, the common banded dotterel, the black-fronted dotterel and the endemic wrybill. Also included among the Charadriiformes are the recently self-introduced Australian spur-winged plover, the endemic black-billed gull and the black-fronted tern.

Although some of the Arctic migrant waders, such as godwits, sandpipers and a few of the plovers, sometimes visit coastal freshwater wetlands, particularly Lake Ellesmere, they are not discussed in this book, as they are essentially marine inhabitants.

SOUTH ISLAND PIED OYSTERCATCHER

Haematopus ostralegus finschi

Abundant • Protected • Length: 46 cm • Maori name: Torea • Status: Native

The South Island pied oystercatcher is a subspecies of the oystercatcher that inhabits many countries worldwide, and the birds are increasing in numbers in New Zealand.

Oystercatchers eat a varied diet, with their long but strong bills enabling them to probe for earthworms, crustaceans and bivalve shellfish. I have noticed a variation in the shape of the

oystercatchers' bills. When feeding and probing for food in soft mud the bill appears to be worn down to a sharp point, but when feeding in rocky terrain for shellfish the tip of the bill becomes blunter with wear.

I studied some oystercatchers feeding in part of the Manukau Harbour, where boat tractors had crushed areas of Pacific oysters. Some birds fed regularly at low tide on these crushed oysters.

Above: Large numbers of South Island pied oystercatchers migrate to North Island marine harbours after nesting.

Left: A South Island pied oystercatcher at its nest in a Canterbury riverbed.

Although I was unable to positively identify individual birds, I noticed over time that birds which had probably recently arrived from feeding in soft habitats in the South Island first had sharply pointed bills, but after a few weeks the sharp tips of some bills had become blunted.

Although oystercatchers are normally able to prise open bivalve shellfish, they appear unable to open oysters. However, in this instance the birds had an easy, regular feed of crushed oysters assisted by a boat-towing tractor.

Very large numbers of the birds inhabit marine harbours in the North Island, particularly in winter months, even though they nest only in the South Island.

Usually their nests are a mere scrape in the shingle of riverbeds, but oystercatchers also nest in pastures and ploughed agricultural land and sometimes high in subalpine locations.

Nesting begins in August and often as late as November, with the clutch of two or three eggs being incubated by each parent in turn for 28 days. The precocial chicks are mobile soon after hatching, remaining with the parents until they disperse in late summer, usually to North Island marine habitats.

AUSTRALASIAN PIED STILT
Himantopus himantopus leucocephalus

Cosmopolitan • Common; widespread
• Protected • Length: 35 cm
• Maori name: Poaka • Status: Native

Australasian pied stilts are cosmopolitan and in some countries they are known as black-winged stilts. As land continues to be cleared, many different feeding habitats have become available for the birds to feed and nest in, and it now appears that their numbers are increasing.

Pied stilts feed by probing for earthworms, crustaceans and insect larvae. Their long legs

A pied stilt feeding along the edge of a lagoon.

enable them to wade in deep water and they will completely submerge their heads to seize prey. In windy conditions I have watched them standing in one position and seizing insects that are blown across the surface of the water towards them. I also noticed that while feeding, a bird would stride to the dry ground at the edge of the water to defecate and then resume feeding.

The graceful, ballet-like movements of the birds are particularly noticeable as they stalk along the banks of wetlands in search of food. I have often been unable to resist taking a surfeit of film with my movie camera when observing their elegant movements as they feed.

While filming pied stilts on a shallow coastal lagoon, I have watched their courtship behaviour from my hide. They were too distant to photograph successfully. However, on several later occasions I was able to photograph the delicate ritual by viewing with binoculars.

As the birds approached each other in the shallow water, the female assumed a posture with body held horizontally and her neck extended. The male then circled her, flicking water at her with his bill. After circling three or four times he mounted her in a brief copulation. Immediately after, the birds just momentarily crossed their long black bills, before separating to feed. I had frequently witnessed this final embrace, but only on one occasion were the birds facing the camera.

In northern districts pied stilts begin nesting in June. Nests are generally located in partially flooded paddocks and lowland marshy wetlands. Although I have found birds nesting in isolation in these habitats, the birds usually form loose colonies. Unfortunately, nests on paddocks are vulnerable as grazing cattle often destroy them.

Most nesting takes place from late winter months to early summer. I have observed these nests situated in a wide variety of habitats: shingle riverbeds, marshy wetlands, cultivated

Top left: A pied stilt at its nest.

Below left to right: The courtship ritual begins with the male circling the female several times, flicking water at her.

Mounting prior to copulation.

The pair mating.

Before parting, the birds momentarily cross bills.

BLACK STILT
Himantopus novaezelandiae

Locally common; highly endangered
• Protected • Length: 40 cm
• Maori name: Kaki • Status: Endemic

The black stilt is one of the rarest wading birds in the world, with the main population being confined to the Waitaki Basin in the South Island, particularly along the relatively unspoilt Ahuriri River. This river has not been exploited for use in the Waitaki hydroelectric scheme.

Up until the 1930s black stilts inhabited wetlands in the Otago region and the lowland swamps of South Canterbury. However, their numbers have drastically declined due to predation by stoats and feral cats.

Today, the Department of Conservation has undertaken a reasonably successful captive breeding programme and intensive predator control to save this rare wader.

Black stilts are larger than the common pied stilt. One problem is that wild birds interbreed with pied stilts, and in winter months some of these hybrids, which display varying amounts of white, are seen mingling with pied stilts in marine harbours of the North Island.

The characteristics of black stilts are quite different from pied stilts. They evolved isolated in New Zealand, possibly millions of years ago. Their habits differ in that they are non-migratory. They do not colonise to nest, and are unfortunately naive towards predators. And as hybrid birds were first observed a century ago, today's black stilts may not be genetically pure, something that could be proved by DNA tests. Immature black stilts do have a black-and-white plumage, but this pattern differs from hybrid colourations.

A three-day-old pied stilt chick.

agricultural land, sheltered shingle beaches, and even single nests on exposed coastal sand dunes.

When humans or other intruders approach nesting sites, both birds leave their nests and, joined by other birds of the colony, dive at the intruder. They frequently feign injury by scraping their wings along the ground in an attempt to lead the intruder away from the nest site. This tactic is often effective when dogs are involved.

Both birds of a pair build a nest composed of a few small sticks or grasses in a depression in the ground, although in wetland habitats the nest is usually built on a high point close to water. Nests on sand dunes are a mere scrape in the sand with little or no addition of nesting material.

Most pied stilt clutches contain two to four eggs, very similar in size and colouration to those of the New Zealand dotterel. Both parents share incubation for 24 or 25 days. As the incubation period is delayed until the full clutch is laid, the chicks hatch within a few hours. The chicks are precocial and covered in pepper coloured down. Although I have seen a young chick feeding alone, they usually stay feeding with the parents for a few months before dispersing.

During the winter months, pied stilts move to coastal lagoons and wetlands, and many settle in sheltered marine harbours and estuaries.

Top right: The highly endangered black stilt.

Bottom right: A hybrid black stilt among South Island pied oystercatchers.

Black stilts feed on insects, crustaceans and other invertebrates. They also take various small fish and probe in soft mud for earthworms and insect larvae.

The birds begin courtship in August and both parents share the task of building a nest in isolation on islands, in shingle riverbeds or on the banks of small tributaries. Up to four eggs are laid, which each parent incubates in turn for 25 or 26 days. The precocial chicks feed and remain with the parents for about six weeks.

Most birds then disperse to feed on wetland lagoons and tarns, and can also be seen around the delta of the Ahuriri River.

BANDED DOTTEREL
Charadrius bicinctus bicinctus

Abundant; widespread • Protected • Length: 20 cm • Maori name: Tuturiwhatu • Status: Endemic

Banded dotterels are found in a very wide range of habitats. They frequently nest on shingle riverbed wetlands and feed around edges of lakes. As well, they feed on and inhabit coastal lagoons, marine harbours, cultivated farmland and sandy beaches. They are also found at high altitudes and regularly nest in the inhospitable Rangipo Desert in the central North Island.

During autumn, many banded dotterels, in particular those that have nested in the South Island, migrate to spend the winter months in Tasmania and some to coastal Australia. However, they return to nest in New Zealand in early spring. Most of the North Island breeding population remains in coastal New Zealand throughout the year.

Mudflats and tidal estuaries are the main feeding grounds for banded dotterels, but many feed on the shores of freshwater wetlands and

Below: A banded dotterel incubating on its riverbed nest.

Right: A male banded dotterel.

Top: Banded dotterels at a riverbed nest.

Above: Banded dotterels feign a 'broken wing' distraction display when their nest is approached.

coastal lagoons, taking a variety of invertebrates. Along stony riverbeds they feed on mayflies, caddis flies and the larvae. On the riverbanks, the birds take earthworms and insects.

Although banded dotterels use a variety of nesting habitats, including sand dunes, short grassy paddocks and agricultural land, their favoured nesting sites are the shingle riverbeds of the South Island and some Hawke's Bay rivers.

Most nests are a scrape in the gravel or sand with a little added decoration of small sticks and sometimes shells. The usual clutch consists of three or four darkly marked eggs that are laid from late winter through to early summer. Incubation, shared by both parents, takes 25 or 26 days, with the hatched chicks being precocial and able to leave the nest within a few hours of hatching. In some instances, however, incubation begins after the laying of the first and second eggs. These earlier hatched chicks remain in the nest until hatching has completed.

Banded dotterels used to regularly nest at Miranda on the Firth of Thames each spring in the short grass near the old limeworks, and here I photographed a pair that had made their nest in the hollow of a dried cow-pat. Banded dotterels feign a 'broken wing' distraction display when intruders approach their nest.

After breeding, the adults moult and lose the distinctive double-band breast markings. They may often be seen feeding together in flocks.

BLACK-FRONTED DOTTEREL
Charadrius melanops

Uncommon; expanding its range
- Protected • Length: 18 cm
- Status: Native

These small Australian dotterels introduced themselves to the shingle riverbeds of southern Hawke's Bay in the early 1950s. Here they bred successfully and have gradually extended their range to other North Island shingle rivers and to some on the east coast of the South Island. After nesting, the birds move in small groups to feed on coastal lagoons, as well as the shores of small lakes, lagoons and estuaries. However, many birds will also remain on the riverbeds.

Below: A black-fronted dotterel approaches its mate incubating eggs on its nest on the bare stones of a Hawke's Bay river.

Bottom: A black-fronted dotterel on a nest camouflaged with driftwood.

The dotterels feed mainly on earthworms and a variety of other invertebrates and their larvae, and aquatic crustaceans.

When disturbed by humans during the nesting season, they crouch, placing their backs to the observer, and are effectively camouflaged among the river stones. Their nests are also extremely difficult to find, with the colours of the eggs blending so well with the surroundings.

The nests are a scrape in the shingle decorated with a few small sticks and often placed near large driftwood. Each parent in turn incubates the three buff-coloured eggs and the precocial chicks hatch after 21 days' incubation.

WRYBILL
Anarhynchus frontalis

Locally common • Protected • Length: 20 cm • Maori name: Ngutuparore • Status: Endemic

The wrybill is unique in the world, in that the end of its bill twists to the right. It is thought that this feature probably enables the birds to more readily capture insects and grubs from beneath rounded river stones on the South Island shingle riverbeds. Walter Buller named the bird the 'crook-bill plover'.

On their shingle nesting habitats along the braided rivers on the eastern side of the Southern Alps, the wrybills' cryptic plumage matches the colour of the riverbed stones. The birds are well camouflaged and, until they move, are extremely difficult to see.

Wrybills feed on a variety of insects as well as their larvae, particularly caddis fly larvae. When rivers are in flood the birds feed along the riverbanks on insects and earthworms.

From August to December wrybills nest on fans of river shingle, devoid of vegetation.

A flock of wrybills in winter, at Miranda in the North Island.

A wrybill at its nest.

Without added nesting material, a scrape is made in the sand between larger stones. The two grey eggs that form the clutch are the same colour as the smooth, rounded stones, disguising them effectively from the eyes of harriers or the occasional marauding black-backed gull.

Incubation, which is shared by the parents, lasts for four weeks, and soon after hatching the precocial chicks learn to find food. Some of the late nests are a result of first clutches having been destroyed by flash floods.

After nesting, most wrybills migrate to the marine harbours of the North Island, particularly the Manukau and Kaipara harbours as well as Miranda on the Firth of Thames. Here their feeding methods change, particularly when they use their bills to sweep from side to side, filtering organisms from the ooze.

Wrybills can be approached very easily when they roost together in large flocks on the marine shell banks, and their aerial flights are a spectacular sight as they twist in unison, with the birds shimmering in the changing light.

SPUR-WINGED PLOVER
Vanellus miles novaehollandiae

Widespread; abundant • Protected • Length: 38 cm • Status: Native

Spur-winged plovers, also known as masked plovers, were self-introduced from Australia in the 1930s. They bred successfully and have spread throughout New Zealand, inhabiting

Top right: Spur-winged plover feeding.

Bottom right: Spur-winged plovers nesting.

open country and farm pastures, particularly where there are ponds or streams. The birds are also seen on the shorelines of lakes, lagoons and on sheltered coasts.

Top: *A plover flock in flight.*

Above: *A spur-winged plover, showing wing-spurs.*

The birds have a distinctive flopping lapwing flight and a high-pitched rattling call. Farmers consider them to be beneficial as they feed on grass grubs and other insect pests. They also eat earthworms and a wide variety of invertebrates, sometimes feeding along the shoreline in coastal habitats and in the shallows of freshwater lagoons, where they also often bathe vigorously.

Plovers have an extended nesting season, from June through to March, with the birds often rearing two or more broods a season. Nests are usually located on rough open pasture or dried riverbeds, being just a scrape in the ground with the addition of small sticks and grasses.

The usual clutch consists of three or four darkly blotched eggs, which are incubated by both sexes in turn over approximately four weeks. Chicks are active soon after hatching and are able to feed for themselves, but remain near their parents for many months.

From my hides I have observed several nests and it seems the birds change 'shifts' (taking turns to incubate) approximately every two hours. But this routine may not apply at night.

Black-billed gull.

BLACK-BILLED GULL
Larus bulleri

> Widespread, mainly on South I. inland waterways; coastal during winter
> • Protected • Length: 37 cm • Maori name: Tarapunga • Status: Endemic

Black-billed gulls differ from other gulls in that they inhabit mainly inland waterways. They are predominantly located on South Island riverbeds and lakes, but they are now also found on some riverbeds in Hawke's Bay and have recently become established on a few other, mainly southern, North Island locations.

During winter, many birds move to coastal locations to feed on molluscs, crustaceans and small fish. In inland areas their food consists of insects, earthworms and other invertebrates and small fish. On agricultural land they may be seen following a plough where worms and grubs have been freshly exposed.

The birds are a slimmer build than red-billed gulls, with a longer, thinner black bill and dark eye rims. Black-billed gulls tend to be less

Top left: A black-billed gull and its nest.

Below left: A black-billed gulls' crèche.

Below: A black-fronted tern in flight.

aggressive and less confiding of humans than other gulls.

Most black-billed gulls nest in the South Island in large colonies on the shores of inland lakes and along braided riverbeds. Other nesting sites have formed in the North Island, notably a thriving colony in the volcanic area of Lake Rotorua around Sulphur Point. Here the common red-billed gull with which they sometimes interbreed joins them. Other smaller North Island colonies have been established recently at Miranda, in the Firth of Thames, and in the Manukau Harbour.

The gulls nest from October to December, building substantial nest mounds with sticks and dried grasses. The usual clutch is two eggs, being incubated by each parent in turn for about 25 days. When the chicks are approximately five days old they congregate in crèches that a few parents take turns to guard. For safety, these are usually situated near the water's edge.

In recent years, some of the nesting colonies on South Island riverbeds have been falling prey at night to feral cats and stoats.

BLACK-FRONTED TERN
Sterna albostriata

Common; widespread, mainly in South I. riverbeds or on coast after nesting
● Protected ● Length: 30 cm ● Maori name: Tarapiroe ● Status: Endemic

The black-fronted tern is found only in New Zealand, with the birds frequenting the riverbeds east of the Southern Alps in the South Island. They breed there but tend to migrate in winter to coastal estuaries, lagoons and harbours, some moving to the North Island.

They are easily distinguished from the Caspian

and white-fronted terns by their smaller size and bright orange feet and bill. Their typical monk-like black cap becomes mottled white in winter after moulting.

Black-fronted terns are dainty feeders, frequently feeding in flocks above the water, rapidly scooping insects or plunging to seize small fish. The birds also hawk (capture) insects across farmland and search for earthworms, beetles and grass grubs. After dispersing to coastal locations in winter, they feed on marine crustaceans and plankton.

The terns perform aerial courtship displays before nesting in loose colonies on the shingle banks of many South Island rivers, sometimes associating with nesting black-billed gulls. Their nests, shallow scrapes in the shingle lined with twigs or grasses, are placed well apart and often in the shelter of boulders or river stones.

The usual clutch of two or three darkly blotched eggs is laid from late September to November. Both parents, taking just over three weeks, share incubation. After hatching, the chicks soon wander about but remain in and around the nest for the first few days. They are able to fly at four and a half weeks old.

Some colony sites are affected by encroaching growth of willows and lupins; nests are vulnerable to flooding as well as predation by stoats, feral cats and harriers.

Above: A black-fronted tern incubating.

Right: A black-fronted tern at its nest.

Order Coraciiformes

Kingfishers

This order includes kingfishers and allies, of which there are many species worldwide.

New Zealand has only one native kingfisher, which is common throughout New Zealand, except Otago and alpine habitats, as well as the Australian kookaburra, introduced to Kawau Island by Sir George Grey, which only inhabits the lower North Auckland region.

The New Zealand kingfisher, *vagans*, is a subspecies of the Australian nominate species.

The kingfisher is included with wetland birds because it is not only present in many habitats but it relies on wetland locations to obtain a variety of food.

NEW ZEALAND KINGFISHER
Halcyon sancta vagans

Common; widespread in many habitats
- Protected • Length: 24 cm
- Maori name: Kotare • Status: Native

Kingfishers are easily distinguished from other native birds, in particular with their bright colouring and general behaviour. They are usually seen perched on a prominent tree, post or power line viewing their surroundings for the possibility of prey. When they fly off they do so with a direct flight and rapid wing-beat to perch on another vantage point.

Wetland habitats provide kingfishers with a broad variety of small fish, tadpoles, insects, crustaceans and freshwater crayfish. They often snatch insects from foliage without landing, or they make lightning-quick dives for small fish in shallow water. Unlike the true diving kingfishers that dive deeply for their restricted diet of fish, these forest kingfishers do not dive in deep water, and their food is more varied.

I have timed kingfisher dives when the birds are catching fish in shallow pools. Depending on the distance of the perch from the water, the rapid dive from the perch, catching a small fish and returning to the perch takes less than three seconds. With a perch close to the water, it may take less than two seconds. From my photographs of kingfisher dives, using a high-speed flash, I have discovered that just before the bird's bill enters the water, the semi-transparent third eyelid, or nictitating membrane, is drawn across the eye to protect it from trauma.

Kingfishers have extremely keen eyesight that enables them to spot prey some distance away. On one occasion a bird was perched on a post close to its nest in a clay bank near a small lake. It suddenly spotted its prey and flew with rapid wing-beats across the lake for nearly 100 metres, and snatched a large dragonfly from foliage on the water surface. It returned to its perch and proceeded to bash the large insect to remove its wings before swallowing it whole.

This perching and bashing behaviour is typical of kingfishers, whether they have captured insects, lizards or fish. It kills the prey and makes it supple for swallowing head-first. Freshwater crayfish are a favourite food. I once watched a kingfisher, which had a nest in a bank bordering a small stream in Marlborough, catch seven freshwater crayfish in the space of two hours and feed them to nesting chicks.

Top right: A kingfisher flying to its nest in a clay bank above a stream.

Bottom: Kingfisher with dragonfly (left) and crayfish (right).

As their diet contains fish bones and chitinous material, kingfishers disgorge pellets of the indigestible material. These can be examined microscopically to identify the food taken.

Kingfishers often bathe in small shallow pools or streams. However, I was once photographing white-faced herons from a hide overlooking a small slow-flowing river, when I noticed a kingfisher momentarily splash into the river with wings outspread and return quickly to the perch and shake its feathers. After a few seconds' pause it repeated the procedure three times, again with wings outspread, and each time returning to the perch to shake the water from its feathers before finally starting to preen.

Kingfishers begin nesting in early summer, calling frequently with a repeated high-pitched 'kek-kek-kek' call. But when a pair begins courtship they converse with a more musical, subdued 'kree-kree-kree' call.

Left: A kingfisher commencing a dive for fish (streamlined).

Above: As the diving kingfisher nears the water, its wings and tail are brought forward in a braking action.

Right: As the kingfisher's bill breaks the water surface the nictitating membranes are drawn across the eyes to provide protection.

I have made a particular study of many kingfisher nests over a period of years, recording behaviour. A vertical clay bank or decaying, standing tree trunk is generally chosen to begin making a tunnel. Both birds perch opposite the selected site and take turns at flying full tilt to chip out a piece of earth or wood. When the hole is deep enough to provide a foothold the birds continue boring in woodpecker fashion.

The final tunnel is usually about 12 to 15

Left: Rising from the water, the membranes still cover the eyes.

Above: The kingfisher shakes water from its feathers before making another dive.

Left: A kingfisher with its catch.

centimetres long and inclines slightly upwards at the end, creating an enlarged chamber for the nest. Using no additional nesting material, the female lays up to six white eggs on the bare floor of the chamber. The female incubates the clutch for approximately 18 days; the male kingfisher occasionally takes over for a short spell when the female leaves the nest to feed.

Kingfishers are reasonably aggressive when protecting their nest, so it is surprising that they appear to be no match for the myna, also a hole nester. Although I have recorded only three instances of mynas ejecting nesting kingfishers, it may be a common event in warmer northern districts where myna populations have increased.

In the first instance that I witnessed, kingfishers were nesting in an old pine tree on the edge of an orchard near my home, when I heard a commotion of kingfisher alarm calls. When I investigated, two mynas were fighting the kingfishers. Although I chased away the mynas, as I approached the site the next morning a myna flew from the nest hole, and the broken, but not eaten, kingfishers' eggs were on the ground below the nest. Although appropriate steps were taken to eliminate the mynas, the kingfishers did not return to reuse the nest.

On another occasion, kingfishers were nesting in a clay bank cutting on a country road. Each time I drove past a kingfisher flew from the nest. However, two weeks later, mynas had taken over the nest, below which I discovered a few fragments of white eggshell.

Kingfisher chicks are naked when first hatched but soon develop spine-like feathers, and as they age they begin to make a peculiar rasping sound, each syllable preceded by a gurgle. When the chicks are well grown this constant noise discloses the situation of the nest, as does the soiling of the tunnel entrance with excrement expelled from the nest by the chicks.

Chicks are fed by both parents, with the birds entering the nest when chicks are young, but as the chicks age they appear at the tunnel entrance to receive food on hearing the parents' approach.

In one instance I became involved in rescuing chicks. A pair of kingfishers was nesting less than 2 metres from the ground in a rotting tree close to a friend's house, and when the chicks were a few weeks old a cat killed both parents. I was told of the tragedy when the chicks had been orphaned for a couple of days and were calling for food continuously. I enlarged the nest hole and extracted five spiny-feathered chicks. I took them home to be cared for by my children, but after a few days of eager feeding on mince meat and earthworms the birds became lethargic. I decided that their diet lacked roughage, so we collected small crabs and a few crickets to replace some of the meat, and within a day the chicks

had regained their vigour. They were eventually taken to Auckland Zoo.

With several nests I have found that chicks fledge when 26 or 27 days old. They take turns at protruding their heads from the tunnel entrance to survey their surroundings. I have waited hours with my tiring finger on the movie camera start button attempting to get a shot of a first flight. A chick will protrude half way out of

A kingfisher feeding a newly fledged chick.

the tunnel as though to fly, but after much hesitation it withdraws and is replaced by another cautious chick.

Generally, the chick will make its first flight of about 50 metres before clumsily landing on a branch where it is immediately joined by the male kingfisher, who feeds it on easily obtained insects every few minutes until the chick gains strength to fly to a higher, safer branch. When another chick eventually joins it, the male feeds and cares for it while the female feeds the less developed chicks in the nest.

I have watched fledged chicks staying with the parents for at least two weeks after fledging, but have not been able to ascertain how long the family remains together.

Order Passeriformes

Passerine Birds

Almost half the birds of the world belong to this order. They are often referred to as 'perching birds', although many birds in other orders are also able to perch.

All birds in this order have four toes, with one pointing backwards, and in no cases are the toes webbed.

Chicks are altricial, being hatched helpless and naked, although some birds in other orders also share this feature.

Of New Zealand wetland birds, only the welcome swallow and the fernbird are included in this order.

WELCOME SWALLOW
Hirundo tahitica neoxena

Abundant; widespread • Protected • Length: 15 cm • Maori name: Warou • Status: Native

Although welcome swallows are not strictly wetland birds, they obtain much of their insect food from freshwater lakes and rely on the mud gathered from the edge of ponds to cement and bind together the grasses of their nests.

The birds are an Australian species that were self-introduced in the 1950s. They are commonly seen throughout New Zealand, inhabiting open country, swamps, rivers and the coast. They prefer to live near various wetlands close to their sources of food. The swallows are often seen in small groups, perched on telephone lines or fences that they use as vantage points and as a means of launching into flight.

Welcome swallows use their long wings and forked tails to aid them in swift, circling flights

A welcome swallow.

Left: A welcome swallow, collecting mud for nest building.

Below: A welcome swallow with chicks at its nest.

to hawk insects above water or in the air. They may even be seen flying low over wet pasture and puddles, capturing grass flies. In winter months, when insect food becomes scarce inland, many swallows migrate to the coast, where they feed on the flies that hover over kelp.

The birds build rather rough, deep, cup-

shaped nests that are composed of dried grasses bound together with mud, lined with feathers and wool and cemented to walls under bridges or ledges in farm buildings as well as in caves. They lay three to five eggs, from August to February, and both parents take part in feeding the chicks.

Above: A welcome swallow flying to its nest.

Top right: A fernbird showing fernlike tail.

Bottom right: The fernbird runs through herbage like a mouse, seldom flying.

FERNBIRD
Bowdleria punctata

- Locally common; widespread • Protected
- Length: 18 cm • Maori name: Matata
- Status: Endemic

The secretive fernbird is infrequently seen. The small birds remain hidden in swamp vegetation where they find adequate food and nesting sites. Uncommon in the drier regions to the east of the Southern Alps, they frequent most wetlands where there is a cover of raupo, reeds and flax.

In addition to freshwater wetlands, fernbirds are found in salt-marshes among sedges and oioi (jointed rush). They also occupy some of the drier northern regions, such as Tongariro National Park, among tussock, flax and bracken.

There are four subspecies of the endemic fernbird in mainland New Zealand, and a fifth subspecies on the subantarctic islands. Fernbirds are unique in that the barbs of their tail feathers are disconnected, resulting in a wispy fern-like appearance to the tails. Their short wings result in a weak flight, and the birds flutter low across vegetation when forced to leave cover.

Fernbirds are insectivorous. When feeding,

they move through the vegetation like mice, searching for beetles, grubs, flies, moths and spiders, as well as the tiny spiderlings that emerge from nursery webs woven on wetland grasses.

The nesting season is extended. In Northland I have found fernbird nests in July and as late as March. Mammalian predators rob a high percentage of nests. Fernbirds build loose and cup-shaped nests of dried grasses, with most clutches consisting of three or four eggs. Both birds incubate these in turn for 14–15 days and the chicks fledge when about 16 days old. After

leaving the nests, the juveniles remain dependent on their parents for a few weeks.
Because of its weak flight and its dependence on wetlands, the fernbird cannot adapt easily to changes in its environment caused by drainage.

Top left: A fernbird in flight.

Bottom left: A fernbird bringing a feather to the nest.

Below: A fernbird bringing a spider to feed chicks.

GLOSSARY

Altricial Hatched without feathers, with eyes closed, and relying totally on parents for warmth and food.
Aquatic Frequenting water.
Brackish water Water containing varying amounts of salt.
Brood To sit over chicks to keep them warm.
Chitin The indigestible material of crustaceans' shells and insect wings.
Cosmopolitan Inhabiting many countries.
Crepuscular Active at dawn and dusk.
Cryptic Colouring which merges with the surroundings, aiding concealment.
Eclipse plumage Dull plumage of males, acquired after breeding.
Endemic Living and breeding in New Zealand, and confined solely to the New Zealand region, e.g. fernbird.
Feral A domesticated animal that has become wild.
Fledged Fully feathered and able to fly.
Hawk To hunt or attempt to capture, especially on the wing.
Introduced Brought into New Zealand from elsewhere, whether self-introduced or by human agency (e.g. mute swan, introduced from Britain in the 1860s).
Invertebrates Animals without a spinal column, e.g. earthworms, insects, spiders.
Juvenile Young bird in its first plumage.
Lamellae Sieve-like appendages on the edges of a bird's bill, used to filter organisms from water or ooze.
Mollusc An animal having an outer shell and soft body, e.g. snails.
Native Naturally occurring in New Zealand, and breeding here, but also found elsewhere in the world, e.g. grey duck, which also inhabits Australia.
Nominate subspecies A subspecies indicated by the repetition of the specific name, e.g. the Australian little grebe, *Tachybaptus novaehollandiae novaehollandiae*, is the nominate subspecies of *Tachybaptus novaehollandiae*.
Precocial Able to display independent activity at birth, especially birds that are hatched covered with down and with open eyes.
Scrape A shallow depression in the ground to be used as a nest.
Speculum The brightly coloured area on a duck's wing.
Subspecies A geographical population of a species, bearing some slight differences from others of the same species.
Threatened (Species that is) declining or low in numbers.
Tibia The shin bone.

BIBLIOGRAPHY

Atlas of Bird Distribution in New Zealand. Wellington: Ornithological Society of New Zealand, 2007.

Buller, Walter L., *Manual of the Birds of New Zealand.* Government Printer, 1882.

Heather, Barrie, and Hugh Robertson, *Field Guide to the Birds of New Zealand* (Second edition). Auckland: Viking Penguin, 2000.

Hosking, Eric, *Wildfowl.* (Text Janet Kear) London: Croom Helm, 1985.

Moon, Geoff, *New Zealand: Land of Birds.* Auckland: New Holland, 2001.

Moon, Geoff, *A Photographic Guide to Birds of New Zealand.* Auckland: New Holland, 2002.

Moon, Lynnette, *Know Your New Zealand Birds.* Auckland: New Holland, 2006.

Turbot, E.G. (Convenor), *Checklist of the Birds of New Zealand.* Third edition, Auckland: Ornithological Society of New Zealand and Random Century, 1990.

INDEX

Anarhynchus frontalis 17, 19, 20, 116–18
Anas aucklandica chlorotis 84–8
    gracilis 83–84
    platyrhynchos platyrhynchos 80–81
    rhynchotis variegata 86
    superciliosa superciliosa 82
Ardea novaehollandiae novaehollandiae 48–55
Aythya novaeseelandiae 11, 13, 86–7

bittern, Australasian 13, 58–63
bogs 10
Botaurus poiciloptilus 13, 58–63
Bowdleria punctata 13, 136–9
Branta canadensis maxima 11, 72–3

Charadrius bicinctus bicinctus 19, 112–15
    melanops 19, 115–16
Circus approximans 19, 88–91
coot, Australian 101–03
coots, introduction 92
cormorant, black 36–8
    pied 38–42
cormorants *see* shags
crake, marsh 13, 98–9
    spotless 95–8
Cygnus atratus 11, 13, 16, 70–71
    olor 68–9

dabchick, New Zealand 13, 16, 27–34
ditches, drainage 10
diurnal birds of prey, introduction 88
dotterel, banded 19, 112–15
    black-fronted 19, 115–16
duck, blue 18–19, 76–9
    grey 82
ducks, introduction 68
dune lakes 10

Egretta alba modesta 16, 55–7

fernbird 13, 136–39
fernbirds, introduction 134
Fulica atra australis 101–03

geese, introduction 68
goose, Canada 11, 72–3
grebe, Australasian crested 16, 24–7
    Australian little 34–5
grebes, introduction 24
gull, black-backed 19, 22
    black-billed 19, 20, 121–3
gulls, introduction 104

Haematopus ostralegus finschi 19, 104–06, 111
Halcyon sancta vagans 9, 126–33
harrier, Australasian 19, 88–91
harriers, introduction 88
heron, white 16, 55–7
    white-faced 48–55
herons, introduction 48
Himantopus himantopus leucocephalus 19, 106–10
    novaezelandiae 20, 22, 110–12
Hirundo tahitica neoxena 134–6
Hymenolaimus malacorhynchos 18–19, 76–9

kahu 19, 88–91
kaki 20, 22, 110–12
karuhiruhi 38–42
kawau 36–8
kawau-paka 44–7
kawau-tui 42–4
kingfisher, New Zealand 9, 126–33
kingfishers, introduction 126
koitareke 13, 98–9
kotare 9, 126–33
kotuku 16, 55–7
kotuku-ngutupapa 16, 64–7
kuruwhengi 86

lagoons, coastal 10
lakes 11–16
    dune 13, 16
    large 11–13
    man-made 13
    subalpine 16
Larus bulleri 19, 20, 121–3
mallard 80–81

matata 13, 136–9
matuku-hurepo 13, 58–63
moho-pereru 92–4

ngutuparore 17, 19, 20, 116–18

oystercatcher, South Island pied 19, 104–06, 111

papango 11, 13, 86–7
parera 82
passerine birds, introduction 134
pateke 84–5
Phalacrocorax carbo novaehollandiae 36–8
    melanoleucos brevirostris 44–7
    sulcirostris 42–4
    varius varius 38–42
Platalea regia 16, 64–7
plover, masked 118–20
    spur-winged 118–20
poaka 19, 106–10
Podiceps cristatus australis 16, 24–7
Poliocephalus rufopectus 13, 16, 27–34
ponds 10–11
Porphyrio porphyrio melanotus 100–01
Porzana pusilla affinis 13, 98–9
    tabuensis plumbea 95–8
pukeko 100–01
putangitangi 74–5
puteketeke 16, 24–7
puweto 95–8

rail, banded 92–4
rails, introduction 92
Rallus philippensis assimilis 92–4
Ramsar convention 9–10
rivers 16–22
    braided 17–22

scaup, New Zealand 11, 13, 86–7
shag, black 36–8
    little 44–7
    little black 42–4
    pied 38–42
shags, introduction 36
shelduck, paradise 74–5

shoveler, New Zealand 86
spoonbill, royal 16, 64–7
Sterna albostriata 19, 123–5
stilt, Australasian pied 19, 106–10
    black 20, 22, 110–12
swallow, welcome 134–6
swallows, introduction 134
swamphen (purple) 100–01
swamphens, introduction 92
swamps 10
swan, black 11, 13, 16, 70–71
    mute 68–9
    white 68–9
swans, introduction 68

Tachybaptus novaehollandiae novaehollandiae 34–5
Tadorna variegata 74–5
tarapiroe 19, 123–5
tarapunga 19, 20, 121–3
teal, black 11, 13, 86–7
    brown 84–5
    grey 83–4
tern, black-fronted 19, 123–5
    Caspian 22
terns, introduction 104
tete 83–4
torea 19, 104–06, 111
tuturiwhatu 19, 112–15

Vanellus miles novaehollandiae 118–20

waders, introduction 104
warou 134–6
wetlands 8–22
    types 10–22
weweia 13, 16, 27–34
whio 18–19, 76–9
wrybill 17, 19, 20, 116–18

Also from New Holland Publishers

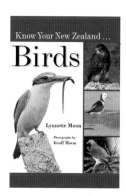

Know Your New Zealand Birds
Lynnette Moon,
photographs Geoff Moon
ISBN 978 1 86966 089 5

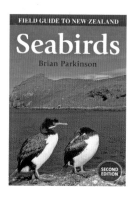

A Field Guide to New Zealand Seabirds
(second edition)
Brian Parkinson
ISBN 978 1 86966 150 2

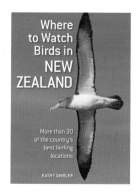

Where to Watch Birds in New Zealand
Kathy Ombler
ISBN 978 1 86966 154 0

Birds of New Zealand Checklist
Brian O'Flaherty
ISBN 978 1 86966 119 9

Photographic Guide to New Zealand Birds
Geoff Moon
ISBN 978 1 877246 58 6